Epidemiologic Methods

The Essentials

Epidemiologic Methods

The Essentials

Stephen C. Newman
University of Alberta
Edmonton, AB, Canada

Academic Press is an imprint of Elsevier
125 London Wall, London EC2Y 5AS, United Kingdom
525 B Street, Suite 1650, San Diego, CA 92101, United States
50 Hampshire Street, 5th Floor, Cambridge, MA 02139, United States
The Boulevard, Langford Lane, Kidlington, Oxford OX5 1GB, United Kingdom

Copyright © 2023 Elsevier Inc. All rights reserved.

No part of this publication may be reproduced or transmitted in any form or by any means, electronic or mechanical, including photocopying, recording, or any information storage and retrieval system, without permission in writing from the publisher. Details on how to seek permission, further information about the Publisher's permissions policies and our arrangements with organizations such as the Copyright Clearance Center and the Copyright Licensing Agency, can be found at our website: www.elsevier.com/permissions.

This book and the individual contributions contained in it are protected under copyright by the Publisher (other than as may be noted herein).

Notices

Knowledge and best practice in this field are constantly changing. As new research and experience broaden our understanding, changes in research methods, professional practices, or medical treatment may become necessary.

Practitioners and researchers must always rely on their own experience and knowledge in evaluating and using any information, methods, compounds, or experiments described herein. In using such information or methods they should be mindful of their own safety and the safety of others, including parties for whom they have a professional responsibility.

To the fullest extent of the law, neither the Publisher nor the authors, contributors, or editors, assume any liability for any injury and/or damage to persons or property as a matter of products liability, negligence or otherwise, or from any use or operation of any methods, products, instructions, or ideas contained in the material herein.

ISBN: 978-0-443-18780-3

For information on all Academic Press publications
visit our website at https://www.elsevier.com/books-and-journals

Publisher: Stacy Masucci
Acquisitions Editor: Elizabeth A. Brown
Editorial Project Manager: Susan Ikeda
Production Project Manager: Punithavathy Govindaradjane
Cover Designer: Vicky Pearson

Typeset by VTeX

To Sandra

Contents

Preface

This book is designed to be a one-semester course in epidemiologic methods at the intermediate level. It is based on my experience teaching epidemiology to graduate students at the University of Alberta.

The book focuses on what I consider to be the core content of epidemiologic methods. Chapter 1 describes several classic studies from the history of epidemiology. These are stories every epidemiologist should know. Chapter 2 covers a range of foundational concepts that appear repeatedly throughout the book, including, the natural history of disease, association and causation, sampling, and study error. Chapters 3 and 4 provide an overview of prevalence studies and cohort studies, respectively. These study designs provide the ideal context in which to introduce the fundamental concepts of prevalence and incidence, which underlie so much of epidemiology. Chapter 5 discusses populations using a hybrid of ideas from epidemiology and demography. Specifically, a population is viewed as an open cohort, and in that setting such traditional approaches to demographic data analysis as direct and indirect standardization are described. Also, the notion of a stationary population is introduced, and the classic Prevalence $=$ Incidence $\times$ Duration equation is presented, nicely tying together core elements of prevalence and cohort studies. Chapters 6–10 discuss effect modification, standardization, bias, and confounding in cohort studies. Chapter 11 covers the basics of randomized trials, which is really just a variation on the cohort theme. Thus, one way or another, Chapters 4–11 are focused on cohort studies. Chapters 12 and 13 discuss case-control studies, adopting the contemporary perspective that a case-control study is essentially a method of sampling from an underlying cohort study. Chapter 14 describes ecologic studies, with an emphasis on the ecologic fallacy. Chapter 15 covers screening in populations, bringing together themes from the earlier discussion of prevalence studies and randomized trials.

Epidemiology is a mathematical discipline, and this is reflected in the way Chapters 2–15 are presented. Various identities and inequalities are found throughout these chapters, but details of proofs and derivations have been relegated to the appendices. The mathematics can be found in the epidemiologic literature, but is included here in a unified manner for the convenience of interested readers. For those not comfortable with mathematical arguments, be reassured that the appendices can be ignored without detracting from the rest of the book.

I devote almost no space to statistical inference. In fact, there is not a single *p*-value or confidence interval within these pages. Most students studying epidemiology at this level will be enrolled in a concurrent course on biostatistics. In my opinion, a discussion of statistical inference would be an unnecessary and potentially confusing duplication of effort.

There are no exercises in this book. That way of engaging students is suitable for an introductory course in epidemiology, but seems out of place here. At this level, students benefit greatly from exposure to the epidemiologic literature. This makes the discipline feel alive and is the next best thing to actually conducting an epidemiologic study. As an alternative to exercises, my recommendation to instructors is to organize small-group sessions in which published papers are discussed and critiqued. I have found this to be an excellent way to reinforce ideas presented in classroom lectures.

The manuscript was typed using the scientific word processor EXP® and the TEX typesetting system MacTeX. The causal diagrams were created using the TEX macro package `diagrams.sty` developed by Paul Taylor. I am pleased to acknowledge the expertise and professionalism of Elizabeth Brown, Punithavathy Govindaradjane, Susan Ikeda, Stalin Viswanathan, and the rest of the Elsevier publishing team. Richard Koch of the University of Oregon kindly provided advice on installing MacTeX. I am indebted to Sander Greenland of UCLA for reviewing portions of Chapters 2, 8 and 13, to George Maldonado and Ali Strickland of the University of Minnesota for reviewing Chapter 8, and to Philip Prorok of the U. S. National Cancer Institute for reviewing Chapter 15. I am further indebted to Sander Greenland for providing valuable insights into various aspects of epidemiologic methods. Needless to say, any remaining errors or deficiencies in the book are solely my responsibility.

I am very interested in receiving your comments, particularly regarding errors or other shortcomings in the book. These can be emailed to me at `stephen.newman@ualberta.ca`. A list of corrections will be posted on the website `https://sites.ualberta.ca/~sn2`.

My wife, Sandra, provided unwavering support and encouragement throughout the writing of the manuscript. It is my pleasure to dedicate this book to her, with love.

CHAPTER

Classic Studies in Epidemiology

1

Epidemiology has made remarkable progress as a field of scientific inquiry over the past fifty years or so, but its roots go back much further than that. In this chapter, a selection of classic studies from the history of epidemiology is presented. This offers a glimpse into the early development of the discipline and provides examples of study designs that we will return to later on.

1.1 Lind Scurvy Trial

Scurvy is a disease resulting from vitamin C deficiency. Primary dietary sources of this essential nutrient are the citrus fruits, such as lemons, limes and oranges. Historically, scurvy was a particular threat to sailors on extended voyages, where fresh fruit was difficult to obtain. Although others had earlier proposed citrus fruit as a treatment for scurvy, James Lind (1716–1794), a Scottish physician with the British Royal Navy, was the first to conduct a clinical study. His hypothesis was that acidifying dietary intake would help prevent the disease.

In his own words, here is a portion of what Lind recorded about his investigation[1(p.149–153)]:

"On the 20th May, 1747, I took twelve patients in the scurvy, on board the Salisbury at sea. Their cases were as similar as I could have them. They all in general had putrid gums, the spots and lassitude, with weakness of their knees. They lay together in one place, being a proper apartment for the sick in the fore-hold; and had one diet common to all, viz., water-gruel sweetened with sugar in the morning; fresh mutton-broth often times for dinner; at other times light puddings, boiled biscuit with sugar, etc.; and for supper, barley and raisins, rice and currents, sago and wine, or the like. Two of these were ordered each a quart of cyder a-day. Two others took twenty five drops of elixir vitriol three times a-day, upon an empty stomach; using a gargle strongly acidulated with it for their mouths. Two others took two spoonfuls of vinegar three times a-day, upon an empty stomach; having their gruels and their other food well acidulated with it, as also the gargle for their mouth. Two of the worst patients, with the tendons in the ham rigid, (a symptom none the rest had), were put under a course of sea-water. Of this they drank half a pint every day, and sometimes more or less as it operated, by way of gentle physic. Two others had each two oranges and one lemon given them every day. These they ate with greediness, at different times,

Epidemiologic Methods. https://doi.org/10.1016/B978-0-44-318780-3.00007-5
Copyright © 2023 Elsevier Inc. All rights reserved.

upon an empty stomach. They continued but six days under this course, having consumed the quantity that could be spared. The two remaining patients, took the bigness of a nutmeg three times a-day, of an electary recommended by an hospital-surgeon, made of garlic, mustard seed, rad. raphan., balsam of Peru, and gum myrrh; using for common drink, barley-water well acidulated with tamarinds; by a decoction of which, with the addition of cremor tartar, they were gently purged three or four times during the course."

"The consequence was, that the most sudden and visible good effects were perceived from the use of the oranges and lemons; one of those who had taken them, being at the end of six days fit for duty. The spots were not indeed at that time quite off his body, nor his gums sound; but without any other medicine, than a gargarism of elixir vitriol, he became quite healthy before we came into Plymouth, which was on the 16th of June. The other was the best recovered of any in his condition; and being now deemed pretty well, was appointed nurse to the rest of the sick."

Lind concluded,

"... I shall here only observe, that the result of all my experiments was, that oranges and lemons were the most effectual remedies for this distemper at sea."

It was over four decades before the British Navy adopted the practice of incorporating lemon or lime juice into the diet of sailors. This led to a dramatic decline in scurvy among the ranks, and incidentally gave rise to the nickname "limey" for a British sailor.

1.2 Semmelweis Puerperal Fever Study

The Vienna Maternity Hospital was a modern teaching hospital in its day. For decades, obstetricians and midwives treated indigent maternity patients on two state-funded hospital wards, called the First Clinic and the Second Clinic. Prior to October 1840, obstetricians and midwives were equally likely to treat patients on either ward, but as of that date obstetricians worked only in the First Clinic, and midwives in the Second Clinic. Puerperal fever is a bacterial infection of the female reproductive tract that occurs following childbirth. In the worst cases it causes overwhelming sepsis and death. Until October 1840 the maternal mortality ratio (defined to be the number of maternal deaths divided by the number of live births in a given calendar year) was broadly the same on the two wards, but during 1841–1846 it was much higher in the First Clinic than in the Second Clinic.[2(p.64,131)]

Ignaz Semmelweis (1818–1865), a Hungarian obstetrician working at the Vienna Maternity Hospital, sought an explanation for this phenomenon. Admissions to the two wards alternated according to days of the week, and so type of maternity patients admitted to the wards was not an answer; neither were atmospheric conditions, seasons, climates, overcrowding, poor ventilation, and other possibilities that he considered. Semmelweis observed, however, that the obstetricians performed autopsies on patients dying of puerperal fever, but the midwives did not. At the time Semmelweis was formulating his ideas, the germ theory of disease had not yet been established. He developed the hypothesis that puerperal fever was being transmitted to maternity patients by "cadaverous particles" that were not removed by ordinary hand washing following an autopsy. In May 1847 he instituted a policy requiring physicians to disinfect their hands with chlorinated lime prior to examining patients. The results were dramatic. Starting in 1848 the maternal mortality ratio in the First Clinic dropped to the level seen in the Second Clinic.[2(p.131)]

Sadly, the story did not end well for Semmelweis. His insights were initially rejected by the medical establishment, which led to a downward spiral in his personal and professional life. This culminated in admission to a lunatic asylum in 1865, where he died two weeks later of septicemia.

1.3 Farr Cholera Study

Cholera is an infection of the small intestine caused by the bacterium *Vibrio cholerae*. In the most severe cases, there is massive watery diarrhea leading to profound dehydration and death, sometimes within hours of onset. Cholera is spread primarily by water and food that has been contaminated with human feces. Prevention of cholera rests on proper sanitation practices and access to clean drinking water. In the era prior to the germ theory of disease, it was widely believed by the medical community that certain afflictions were caused by invisible organic particles called miasms. According to the miasma theory, these particles were released into the air by decomposing organic matter and subsequently inhaled by the unsuspecting recipient.

William Farr (1807–1883) was an English physician, widely regarded as one of the founders of medical statistics. While serving as the Statistical Superintendent of the General Register Office, he conducted a study of the London cholera epidemic of 1849 that gave results strikingly consistent with the miasma theory.[3] Farr grouped the 38 districts of London according to their mean elevation above the high water mark of the Thames River and determined for each the corresponding cholera mortality rate (defined to be the number of deaths due to cholera in a given calendar year divided by the number of individuals in the population at or near the midpoint of the year). Fig. 1.1 shows the cholera mortality rates in the London districts as a function of mean elevation above the Thames River. Farr reasoned that the lower the elevation, the more the ground would be fouled by sewage and other noxious substances due to runoff from above, leading to progressively more contaminated soil at lower elevations and increasingly unhealthy vapors. The inverse relationship between cholera mortality rate and mean elevation is evident.

1.4 Snow Cholera Study

John Snow (1813–1858) was an English physician, best known in the history of medicine as a pioneer in the development of anesthesia. Among epidemiologists he is renowned for conducting a classic study supporting the germ theory hypothesis

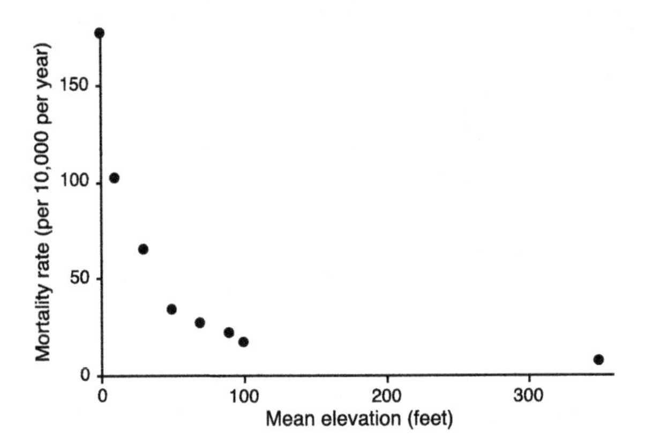

FIGURE 1.1

Cholera mortality rate and mean elevation above the Thames River in London districts, 1849.
Data from Reference 3.

that cholera is a waterborne disease transmitted between humans by the fecal-oral route. The cholera epidemic of 1849, mentioned in connection with John Farr, returned to London in 1853, and again in 1854. Snow saw an opportunity to conduct what we now call a *natural experiment*. In that era, London drinking water was provided to individual residences by private companies drawing their supply from the Thames River. Two such companies were the Southwark and Vauxhall Company and the Lambeth Company. Snow learned that the Southwark and Vauxhall Company obtained water from a part of the Thames contaminated with sewage, while the Lambeth Company accessed its supply from a portion of the Thames free of sewage. Importantly for Snow, both companies had clients in a certain area of London. This offered a unique opportunity for a study of the transmission of cholera.

In his own words, here are some of the key features of the study Snow conducted[4(p.74–75)]:

"*... the intermixing of the water supply of the Southwark and Vauxhall Company with that of the Lambeth Company, over an extensive part of London, admitted of the subject [whether cholera was waterborne or not] being sifted in such a way as to yield the most incontrovertible proof on one side or the other. In the sub-districts ... supplied by both Companies, the mixing of the supply is of the most intimate kind. The pipes of each Company go down all the streets, and into nearly all the courts and alleys. A few houses are supplied by one Company and a few by the other, according to the decision of the owner or occupier at that time when the Water Companies were in active competition. In many cases a single house has a supply different from that on either side. Each Company supplies both rich and poor, both large houses and small; there is no difference either in the condition or occupation of the persons receiving the water of different Companies. ... As there is no difference whatever, either in the houses or the people receiving the supply of the two Water Companies, or in any of the* *physical conditions with which they are surrounded, it is obvious that no experiment could have been devised which would more thoroughly test the effect of water supply on the progress of cholera than this, which circumstances placed ready made before the observer."*

"The experiment, too, was on the grandest scale. No fewer than three hundred thousand people of both sexes, of every age and occupation, and of every rank and station, from gentlefolks down to the very poor, were divided into two groups without their choice, and, in most cases, without their knowledge; one group being supplied with water containing the sewage of London, and, amongst it, whatever might have come from the cholera patients, the other group having water quite free from such impurity."

"To turn this grand experiment to account, all that was required was to learn the supply of water to each individual house where a fatal attack of cholera might occur. ..."

The key findings of this impressive undertaking are summarized in Table 1.1, where the cholera mortality rate for clients of the Southwark and Vauxhall Company is seen to be almost six times as large as the corresponding rate for the Lambeth Company. In the same year that the Snow monograph was published, a reviewer pointed out that the denominators appearing in Table 1.1 are not specific to the area of London where the clients served by the water companies were intermingled, thereby casting doubt on the study findings.[5] Subsequent analyses using the appropriate denominators yielded essentially the same results, thus vindicating Snow and supporting his theory that cholera is (somehow) caused by fecal-oral transmission.[6,7]

Table 1.1 Number of cholera deaths during July to October 1854, estimated population served in 1851, and cholera mortality rates, by water company.

Water company	Deaths	Population	Rate[a]
Southwark and Vauxhall	4,093	266,516	153.6
Lambeth	461	173,748	26.5

[a] *per 10,000 per 4 months.*
Data from Reference 4.

1.5 Lane-Claypon Breast Cancer Study

Janet Lane-Claypon (1877–1967), an English physician and physiologist, conducted a landmark study on the "antecedent histories" of (female) breast cancer patients.[8] Two groups of subjects were recruited from eight hospitals in England and three hospitals in Scotland. The first group, referred to as *cases*, were patients "who either had suffered or were suffering from cancer of the breast". The second group, referred to as *controls*, were "women over 45 who were either in hospital for some trouble, other than cancer, either surgical or medical, or who were applying for out-patient treatment". Cases were restricted to those able to attend their treating hospital for an interview, and only cases and controls living within 20 miles of the hospital were eligible. A questionnaire was administered to cases and controls that included sections on demographics, menstrual history, reproductive history, lactation history and family medical history. For cases, there was also a component on the progression of their breast cancer, as well as a section abstracted from hospital files on clinical presentation and pathology.

There were 508 cases and 509 controls. The report produced by Lane-Claypon details a series of careful analyses. For example, after accounting for age at marriage and duration of marriage using linear regression, cases had 22% fewer children than controls. This finding is consistent with our current understanding of the way pregnancy affects the lifetime risk of breast cancer.

1.6 Doll and Hill Lung Cancer Study

There was a dramatic and unexplained increase in lung cancer deaths in several countries between World War I and World War II. During that era, controversy raged about whether smoking was the underlying cause. Richard Doll (1912–2005), a British physician and epidemiologist, and Austin Bradford Hill (1897–1991), a British statistician and epidemiologist, conducted a landmark study of smoking and lung cancer.[9,10] The design was broadly similar to that employed in the Lane-Claypon breast cancer study. During April 1948 to February 1952, male and female lung cancer patients less than 75 years old were recruited from 20 London hospitals. As each case patient was enrolled, a corresponding control patient with a noncancer diagnosis was recruited from the same hospital, matched on age group, gender and date of admission. Each subject was interviewed using a "set questionary", which included a detailed smoking history. A "smoker" was defined to be "a person who had smoked as much as one cigarette a day for as long as one year".

There were 1,465 cases and an equal number of controls. The proportion of cases who were smokers was 96.8%, and for controls it was 91.8%. This analysis and others based on such characteristics as "most recent amount of tobacco smoked regularly before the onset of the present illness" and "average amount of tobacco smoked daily over the 10 years preceding the onset of the present illness" provide evidence of an association between smoking and lung cancer.

1.7 National Health Survey

The National Health Survey (NHS) of 1935–1936 was conducted by the U.S. government to obtain an overview of health and disability in the country.[11,12] The NHS was a composite of urban and rural surveys; here we focus on the urban component. Data were collected on 2,502,391 citizens living in 703,092 households distributed across 83 cities. The cities were chosen to be representative of the demographic make-up of four geographic regions of the country: northeast, north central, south, west. Within each city, groupings of city blocks, previously created for the 1930 census, were selected and households within each grouping were approached.

The NHS received extraordinary support from the public. Fully 98.5% of households asked to participate agreed to do so. The questionnaire documented the health status of each member of the household at the time of the interview and during the preceding 12 months. Information was collected on demographics, acute and chronic diseases, physical impairments, and medical services received. The NHS found that 4.4% of respondents had a disability at the time of interview, and 18.0% of respondents had a disability lasting one week or longer during the 12 months preceding the interview, where "disability" was defined to be "inability to work, attend school, care for home, or carry on other usual pursuits by reason of disease, accident, or physical or mental impairment".

1.8 MRC Streptomycin Trial

The Medical Research Council (MRC) in the United Kingdom conducted a study, now referred to as a *randomized clinical trial*, to investigate the effectiveness of the newly-developed antibiotic streptomycin in the treatment of pulmonary tuberculosis.[13] To be eligible, subjects had to be 15–30 years old with acute progressive bilateral pulmonary tuberculosis of recent origin that was unsuitable for collapse therapy. For such patients the only accepted treatment at that time was bed rest. During January to September 1947, a total of 107 patients were recruited for the study from England, Scotland and Wales and admitted to a designated hospital for treatment.

An important feature of the study design was a formal randomization scheme using sealed envelopes that assigned subjects to treatment with either streptomycin (plus bed rest) or bed rest (only), and in such a way that treatment assignment was unknown in advance to either the patients or the investigators. Randomization resulted in 55 patients in the streptomycin group and 52 patients in the bed rest group. Patients were not told before admission that they were being enrolled in a clinical trial and might receive an innovative treatment. Streptomycin was administered by intramuscular injection, and so patients in the streptomycin group knew they were receiving a treatment in addition to bed rest. Follow-up of all patients continued in hospital for six months. Chest X-rays were performed on a monthly basis and interpreted by three specialists unaware of treatment assignments. The proportion of patients in the streptomycin group showing radiologic improvement at six months was 69%, and in the bed rest group it was 33%, demonstrating a clear benefit from streptomycin.

1.9 Salk Polio Vaccine Trial

Poliomyelitis, commonly known as polio, is an infectious disease caused by the poliovirus. The vast majority of those who get infected have no symptoms, most of the rest have a flu-like illness, but in a small minority of cases the central nervous system becomes involved, leading to muscle weakness and paralysis, and sometimes death. The National Foundation for Infantile Paralysis (NFIP) in the United States was a community-based nonprofit organization devoted to research and treatment of polio. It was created in 1938 by President Franklin D. Roosevelt, who was himself a victim of polio, having contracted the disease when he was 39 years old, leaving him paralyzed in both legs. In 1952, Jonas Salk (1914–1995), an American physician and virologist, created a polio vaccine. The NFIP sponsored a massive community-based study to determine the efficacy of the Salk vaccine.[14,15]

The Salk vaccine trial was actually a composite of two separate trials, referred to as the placebo control trial and the observed control trial. Here we focus on the placebo control trial, which was conducted in 84 areas of 11 states that were selected on the basis of consistently high rates of polio in the preceding five years. With parental consent, children in grades 1–3 were randomly assigned to receive a series of three injections of either vaccine or placebo. Treatment assignment was unknown to the children, their parents and those performing the vaccinations. Injections were administered during April to June 1954, with 95.9% of children receiving a complete series. For the 401,974 children who received all three injections, the attack rate of paralytic polio (defined to be the number of cases of paralytic polio divided by the number of children under observation) was 16.4×10^{-5} for the vaccinated group and 57.2×10^{-5} for the unvaccinated group, providing convincing evidence of the value of the Salk vaccine.

1.10 Newburgh-Kingston Fluoride Trial

The origins of water fluoridation in the United States began with the observation in 1901 by Frederick McKay (1874–1959), a Colorado dentist, that patients with an unusual brown mottling of their permanent teeth were resistant to dental caries. McKay spent years investigating what was then colorfully termed the Colorado brown stain. In 1931 laboratory scientists determined that the problem was due to excessive fluoride in community water supplies. Further research suggested that a fluoride concentration of approximately 1 part per million (ppm) in the drinking water during the years of tooth development had the potential to provide protection against dental caries with a negligible risk of mottling. To investigate this more closely, a study of the impact of fluoride on the dental health of children was conducted in Newburgh and Kingston, two cities in the state of New York, each with a population of about 30,000.[16,17]

During 1944–1946, baseline dental examinations were performed on all school children 6–12 years old in Newburgh and Kingston. The extent of dental caries in each child was quantified by counting the number of decayed, missing or filled (DMF) teeth. Mean DMF counts in the two cities were similar at baseline, as were the natural fluoride concentrations in the community water supplies. Beginning in May 1945, the Newburgh water supply was supplemented with fluoride to bring its concentration to 1.0–1.2 ppm for the duration of the study, while the Kingston water supply was left untouched. Dental examinations performed after ten years on samples of 1,628 school children from Newburgh and 2,140 school children from Kingston found that the mean DMF count in Newburgh was approximately half that in Kingston, demonstrating the value of supplementing the community water supply with fluoride.

1.11 Framingham Heart Study

In the 1940s, cardiovascular disease was the leading killer in the United States, but its causes were poorly understood and few treatments were available. The Framingham Heart Study (FHS) was undertaken with the aim of uncovering risk factors for cardiovascular disease.[18] (Interestingly, the term *risk factor*, now ubiquitous in epidemiology, was introduced into the lexicon by the FHS investigators.) The town of Framingham, Massachusetts was chosen as the site for the study because its population of about 28,000 was stable, and because both the residents of the town and its medical community were enthusiastic about being involved.

The first group of participants was recruited during 1948–1952 and consisted of 5,209 individuals 28–62 years old without a history of heart attack or stroke; more than half were women. Subjects provided an extensive medical history and underwent a thorough medical assessment, including a physical examination and laboratory tests. Assessment of medical status was repeated approximately every two years, and there was routine surveillance of local hospitals to gather information on cardiovascular events.

The FHS was eventually expanded to include the offspring of the original study participants, and their offspring in turn. As of writing, the FHS is ongoing, making it one of the longest-running studies of its kind. The FHS has produced, and continues to produce, a wealth of information on cardiovascular disease. In particular, it has helped to establish that elevated blood pressure, elevated cholesterol, diabetes and cigarette smoking are risk factors for developing coronary heart disease.[19]

References

1. Lind J. *A treatise on the scurvy. In three parts*. 2nd ed. London: Millar; 1757.
2. Semmelweis I. *The etiology, concept, and prophylaxis of childbed fever*. [Carter KC, trans.] Madison: University of Wisconsin Press; 1983. [Original work published 1861].
3. Farr W. Influence of elevation on the fatality of cholera. *J Stat Soc London* 1852;**15**(2):155–83.
4. Snow J. *On the mode of communication of cholera*. 2nd ed. London: Churchill; 1855. Reprinted in *Snow on cholera*. New York: Commonwealth Fund; 1936.
5. Parkes EA. Review: Mode of communication of cholera. By John Snow, MD. Second edition - London, 1855. *Br Foreign Med Chir Rev* 1855;**15**:449–56. Reprinted *Int J Epidemiol* 2013;**42**(6):1543–52.

6. Koch T. Commentary: Nobody loves a critic: Edmund A Parkes and John Snow's cholera. *Int J Epidemiol* 2013;**42**(6):1553–9.
7. Snow J. Cholera and the water supply in the south districts of London, in 1854. *J Public Health Sanit Rev* 1856;**2**(7):239–57.
8. Lane-Claypon JE. *A further report on cancer of the breast, with special reference to its associated antecedent conditions. Reports on public health and medical subjects, no. 32.* London: His Majesty's Stationery Office; 1926.
9. Doll R, Hill AB. Smoking and carcinoma of the lung: preliminary report. *Br Med J* 1950 Sep 30;**2**(4682):739–48.
10. Doll R, Hill AB. A study of the aetiology of carcinoma of the lung. *Br Med J* 1952 Dec 13;**2**(4797):1271–86.
11. Perrott G, Tibbitts C, Britten RH. The National Health Survey: scope and method of the nation-wide canvass of sickness in relation to its social and economic setting. *Public Health Rep* 1939;**54**(37):1663–87.
12. Britten RH, Collins SD, Fitzgerald JS. The National Health Survey: some general findings as to disease, accidents, and impairments in urban areas. *Public Health Rep* 1940;**55**(11):444–70.
13. Streptomycin in Tuberculosis Trials Committee. Streptomycin treatment of pulmonary tuberculosis: a Medical Research Council investigation. *Br Med J* 1948 Oct 30;**2**(4582):769–82.
14. Francis Jr T, Korns RF. Evaluation of 1954 field trial of poliomyelitis vaccine: synopsis of summary report. *Am J Med Sci* 1955;**229**(6):603–12.
15. Monto AS. Francis field trial of inactivated poliomyelitis vaccine: background and lessons for today. *Epidemiol Rev* 1999;**21**(1):7–23.
16. Ast DB, Schlesinger ER. The conclusion of a ten-year study of water fluoridation. *Am J Public Health Nations Health* 1956;**46**(3):265–71.
17. Ast DB, Smith DJ, Wachs B, et al. Newburgh-Kingston caries-fluorine study. XIV: combined clinical and roentgenographic dental findings after ten years of fluoride experience. *J Am Dent Assoc* 1956;**52**(3):314–25.
18. Mahmood SS, Levy D, Vasan RS, et al. The Framingham Heart Study and the epidemiology of cardiovascular disease: a historical perspective. *Lancet* 2014 Mar 15;**383**(9921):999–1008.
19. Kannel WB. The Framingham Study: its 50-year legacy and future promise. *J Atheroscler Thromb* 2000;**6**(2):60–6.

CHAPTER

Preliminaries

2

In this chapter, a number of basic concepts are presented that appear repeatedly throughout the book.

2.1 Natural History of Disease

It is usual to think of a disease in dichotomous terms—a person either has the disease or not. Throughout this book we instead view a disease as an evolving process punctuated by milestones. The entirety of this evolution is referred to as the *natural history* of the disease.[1–3] Fig. 2.1 depicts the natural history of a disease that can be caused by a certain exposure. Not all diseases fit this paradigm, but it is sufficiently general for our purposes.

Let us consider a concrete example of Fig. 2.1, where the exposure is cigarette smoking and the disease is lung cancer. Once a person starts smoking (*onset of exposure*), a minimum amount of time is needed for the carcinogens in cigarette smoke to produce the characteristic changes in lung tissue (*onset of pathology*) that have the potential to develop into symptomatic lung cancer. Precisely what constitutes the onset of pathology is a matter of definition. Is it, for instance, the presence of a single cancerous cell, or perhaps a tumor of a given size?

The interval from the onset of exposure to the onset of pathology is called the *induction period*. If the individual ceases smoking during the induction period, lung cancer might be avoided. At some point, the disease process reaches a stage where,

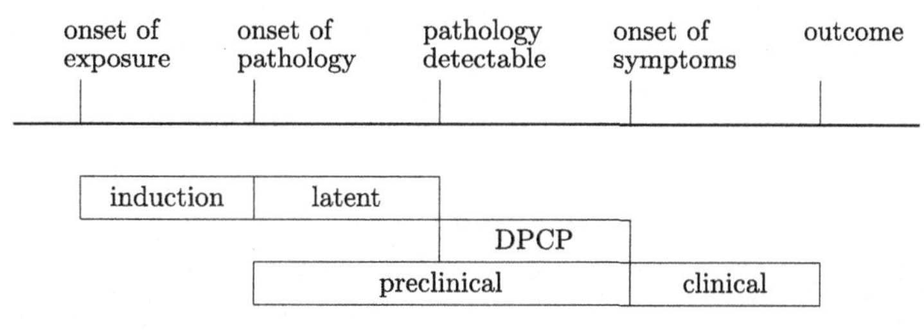

FIGURE 2.1

Natural history of disease.

Epidemiologic Methods. https://doi.org/10.1016/B978-0-44-318780-3.00008-7
Copyright © 2023 Elsevier Inc. All rights reserved.

although not yet causing symptoms, it can be detected with current medical technology (*pathology detectable*). The interval between the onset of pathology and the point where the pathology is detectable is called the *latent period*. As the disease progresses, it eventually starts to cause symptoms (*onset of symptoms*). The interval from the onset of pathology to the onset of symptoms is called the *preclinical phase*. Strictly speaking, it is the disease that is in the preclinical phase, but it is sometime convenient to say that the individual is in the preclinical phase, and likewise for other phases. The interval from the point where pathology is detectable to the onset of symptoms is that portion of the preclinical phase called the *detectable preclinical phase* (DPCP).

At some point after symptoms appear, the individual seeks medical attention and definitive investigations are performed, which we refer to as a *diagnostic assessment*. Here it is assumed that the diagnostic assessment takes place immediately after symptoms are noticed. This evidently does not always hold in practice, but the resulting simplification in Fig. 2.1 is convenient. For suspected lung cancer, a diagnostic assessment might involve, among other things, imaging studies and a tissue biopsy. It is possible for a diagnostic assessment to take place prior to the onset of symptoms. For example, asymptomatic lung cancer might be detected if a suspicious lesion is identified on a chest X-ray performed for reasons unrelated to a possible malignancy. Diagnostic assessments performed on asymptomatic individuals will be discussed in Chapter 15 in the context of community screening programs. Once the diagnosis is made, the disease is considered to be in the *clinical phase*. The disease process eventually culminates in either *recovery* or *death* (from the given disease or some other cause), concluding that *episode* of the disease. Depending on the disease, individuals who recover might face the prospect of further episodes.

Fig. 2.1 gives the appearance of an inexorable process where a disease in the preclinical phase eventually progresses to the clinical phase. However, the natural history need not follow this trajectory. For instance, the disease might simply resolve during the preclinical phase. Alternatively, someone in the preclinical phase of the disease could die of some other condition prior to the disease of interest being diagnosed. This is more likely to occur when the disease has a long preclinical phase, that is, when it is slowly progressive. For example, prostate cancer can remain asymptomatic for years and may be diagnosed postmortem in men dying of other causes.

For purposes of conducting epidemiologic studies, we need to construct a dichotomous version of "disease". We say that someone identified as having a disease on the basis of a diagnostic assessment is a *case* (of the disease), otherwise a *noncase*. According to this definition, case status at any given moment is completely determined by whether there has been a diagnostic assessment and, if so, when it took place. Not all diagnostic assessments are created equal. We assume that each individual has a true case status at the time a diagnostic assessment is performed and that a diagnostic assessment is conducted without error. Such an error-free diagnostic assessment is said to be a *gold standard* for deciding case status.

2.2 Association and Causation

It is widely-accepted that (cigarette) smoking is a "cause" of lung cancer. But what exactly is meant by such an assertion? A correlation exists between smoking and lung cancer, but it is not perfect. There are lifelong smokers who do not develop lung cancer and, on the other side, there are committed nonsmokers who do. For this reason, we are unable to say in advance who will eventually be stricken with this disease. Our uncertainty can be expressed by saying that smoking increases the "risk" of developing lung cancer, or more precisely that smoking increases the probability of becoming a case of lung cancer, where the probability is specified in terms of a particular probability model.

We can quantify the impact of smoking on lung cancer by conducting a study that compares the probability of becoming a case of lung cancer in smokers to the corresponding probability in nonsmokers. A way of formalizing the comparison would be to compute the ratio given by the probability in smokers divided by the probability in nonsmokers. Our proposed study would almost certainly find a ratio greater than 1, which would be evidence of an *association* between smoking and lung cancer. But association is a far cry from *causation*, and a ratio of probabilities greater than 1 is not sufficient to establish the latter. To take an extreme example, suppose that in our study, rather than label subjects according to whether they smoke or not, we instead focus on if they usually carry a pack of cigarettes or not. The ratio of probabilities from this study would be similar, perhaps identical, to the ratio from the earlier study, and would establish an association between carrying a pack of cigarettes and lung cancer. But it is hard to imagine anybody claiming that the act of carrying a pack of cigarettes in and of itself causes lung cancer.

Evidently, causation is a much deeper concept than association, and far more difficult to establish. Epidemiologic studies are well equipped to uncover associations, but rarely, if ever, up to the task of determining causation. That usually requires a multidisciplinary collaboration involving the expertise of epidemiologists, basic scientists, clinicians, statisticians, and other specialists.

It would be ideal if there were a set of guidelines to help with the decision as to whether a given association is causal. A proposal along these lines was put forward in 1965 by Austin Bradford Hill, whose name appeared in connection with the lung cancer study discussed in Section 1.6.[4] At least some of the nine so-called *Hill guidelines* usually find their way into books on epidemiologic methods. Although it can be argued that most of the guidelines have limited practical applicability, two are of enduring importance. *Temporality* specifies that "exposure" precedes "disease" and is an absolute prerequisite for causality. *Biologic plausibility*, which calls for a mechanism that explains how an "exposure" might be a cause of a "disease", is not a requirement for causality, but its absence undermines the credibility of any assertion regarding causation.

Although the focus in much epidemiologic research is on causation, that sentiment is not always reflected in the way studies are reported.[5] Given the difficulties surrounding causality, there tends to be a reluctance in epidemiologic discourse to assert that an exposure is a cause of a disease. Often the language used is more tentative, stating for instance that an exposure is a *risk factor* for a disease. A risk factor might very well be a cause of a disease, but it also includes the possibility that only a noncausal association is present.

2.3 Types of Epidemiologic Studies

As depicted in Fig. 2.2, epidemiologic studies fall into two broad categories—descriptive and etiologic—which are in turn subdivided into more specific types of studies. Table 2.1 lists the studies presented in Chapter 1 along with their designs. Detailed discussion of the various epidemiologic study designs occupies much of the rest of this book.

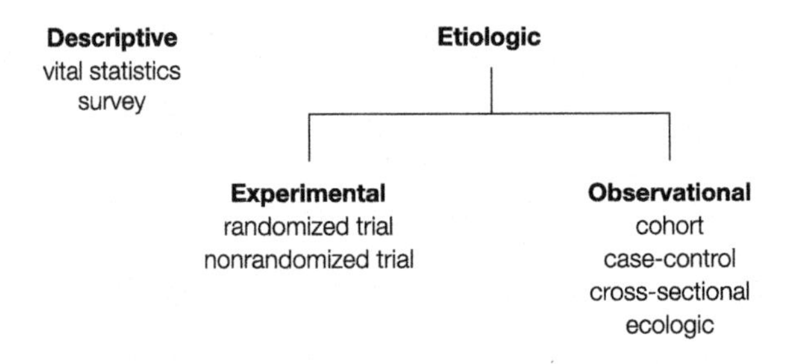

FIGURE 2.2

Types of epidemiologic studies.

Descriptive studies in epidemiology are concerned with characterizing the health status of groups of individuals with respect to person, place and time. These studies address questions such as: what proportion of a population has a given disease?, does the proportion vary over demographic subgroups or geographic regions?, and does the proportion exhibit a trend over time? Studies based on *vital statistics* and *surveys* are types of descriptive studies commonly undertaken by government agencies. They can be specialized undertakings conducted as one-time events or part of an ongoing program of data collection designed to create a health profile of the population over time. Such studies play an important role in monitoring population wellbeing and planning health services. The insights provided by descriptive studies can also be a source of hypotheses to be addressed by more focused etiologic studies. The National Health Survey is a descriptive study that was conducted to provide an overview of health and disability in the U.S. population at a particular time.

Etiologic studies in epidemiology are concerned with investigating causation. These studies address such questions as: does a particular lifestyle increase mortality?, does a newly-developed drug cure a certain disease?, and does a given occupational exposure lead to disability? Etiologic studies are usually subdivided into two categories—experimental and observational.

Table 2.1 Epidemiologic studies presented in Chapter 1.

Study	Design
Lind scurvy trial	nonrandomized trial
Semmelweis puerperal fever study	cohort study
Farr cholera study	ecologic study
Snow cholera study	cohort study
Lane-Claypon breast cancer study	case-control study
Doll and Hill lung cancer study	case-control study
National Health Survey	survey
MRC streptomycin trial	randomized trial
Salk polio vaccine trial	randomized trial
Newburgh-Kingston fluoride trial	nonrandomized trial
Framingham Heart Study	cohort study

Experimental studies in epidemiology are those in which the investigator has a degree of influence over subjects and their exposures, best exemplified by the *randomized trial*. In this type of study, the investigator randomly assigns each subject to one of several treatments. This brings the randomized trial as close to a controlled laboratory experiment as is ethically possible in human populations, giving this study design the best opportunity of any in epidemiology to establish causation. The MRC streptomycin trial and the placebo control component of the Salk polio vaccine trial are examples of randomized trials. Not all experimental studies in epidemiology involve randomization. The Lind scurvy trial and the Newburgh-Kingston fluoride trial are examples of nonrandomized trials. An experimental study in epidemiology that is not randomized is sometimes called a *quasi-experimental study*.

Observational studies in epidemiology encompass a range of study designs, including, *cohort studies*, *case-control studies*, *cross-sectional studies* and *ecologic studies*. In an observational study, the investigator is merely a passive observer of subjects and their exposures, without the ability to intervene. For this reason, it is usually much more difficult to make an argument for causation based on an observational study than on an experimental study. Among observational studies, it is cohort studies that generally provide the strongest evidence for causation, case-control studies perhaps less so, with cross-sectional and ecologic studies typically supplying inferior evidence. The Semmelweis, Farr, Snow, Lane-Claypon, Doll and Hill, and Framingham studies are examples of observational studies.

Distinguishing experimental studies from etiologic studies is usually straightforward, but deciding whether an etiologic study is descriptive or observational can sometimes be difficult. The problem is that descriptive studies frequently include analyses directed at causation, or at least association. For example, the National Health Survey looked at disability status according to whether government "relief" payments were received. This could be regarded simply as a detailed descriptive analysis, but it also casts some light on whether having a disability is associated with financial difficulties.

The analysis of data from descriptive studies usually relies on a range of descriptive statistics such as means and proportions, often presented according to sociodemographic and other categories chosen to reflect the aims of the study. By contrast, the analysis of data from etiologic studies typically involves one or more indices that quantify the extent to which a given exposure is associated with a disease of interest. An index of this type is called a *measure of association*. When the explicit or implicit aim of the study is to investigate possible causation, the index is referred to as a *measure of effect*. However, such distinctions are not always easily made.

From Table 1.1 for the Snow cholera study, the cholera mortality rates were 153.6 and 25.6 per 10,000 per 4 months for clients of the Southwark and Vauxhall Company and the Lambeth Company, respectively. The ratio $153.6/26.5 = 5.80$ shows that the mortality rate for customers of the Southwark and Vauxhall Company was about six times as large as that for clients of the Lambeth Company. Due to the study design, an argument can be made that tainted drinking water (somehow) causes an increase in cholera mortality. On these grounds, it seems reasonable to refer to the ratio of cholera mortality rates as a measure of effect.

2.4 Weighted Averages

For a given positive integer n, let $r_1, r_2, \ldots, r_n$ be arbitrary real numbers, let $w_1, w_2, \ldots, w_n$ be arbitrary nonnegative real numbers, not all of which are zero, and let $W = \sum_{i=1}^{n} w_i$. In the present context, we refer to $w_1, w_2, \ldots, w_n$ as *weights*. The *weighted average* of $r_1, r_2, \ldots, r_n$ is defined to be

$$\begin{aligned} \overline{r} &= \frac{1}{W} \sum_{i=1}^{n} w_i r_i \\ &= \sum_{i=1}^{n} \left(\frac{w_i}{W} \right) r_i , \end{aligned} \tag{2.1}$$

where we note that

$$\sum_{i=1}^{n} \frac{w_i}{W} = 1 . \tag{2.2}$$

We see from (2.1) that the greater the weight w_i assigned to a given r_i, the greater is its contribution to the weighted average $\overline{r}$. Weighted averages have some of the properties expected of "averages":

1. Let $r_{\min}$ and $r_{\max}$ be the minimum and maximum values of $r_1, r_2, \ldots, r_n$. It is shown in Appendix A that $r_{\min} \le \overline{r} \le r_{\max}$. That is, $\overline{r}$ falls within the range spanned by $r_1, r_2, \ldots, r_n$.

2. If all the weights are equal, then $w_i / \sum_{i=1}^{n} w_i = 1/n$ and

$$\overline{r} = \frac{1}{n} \sum_{i=1}^{n} r_i ,$$

which is the usual arithmetic mean.

3. If $r_1, r_2, \ldots, r_n$ are all equal and their common value is denoted by r, then $\overline{r} = r$.

2.5 Sampling

Sampling in epidemiology, as in other areas of research, refers to the process of selecting individuals from a "population" for participation in a study. The group of subjects recruited for this purpose is called a *sample*. We consider three approaches to sampling—probability sampling, nonprobability sampling and random sampling.

Probability Sampling

In *probability sampling*, subjects are selected on a probability basis from a *finite population*, which for the moment we take to mean a geographically-defined group of individuals, such as the residents of a city or a country. It might be possible to collect data on everyone in a finite population, but aside from a government-sponsored *census*, such an undertaking is usually impractical. An alternative is to conduct a *survey* in which information is gathered on a probability sample.[6] The attraction of a probability sample is that it is more economical than a census and yet is *representative* of the population. The term representative is not used here to mean that the probability sample is simply a mirror of the population in miniature, but rather that with the aid of a statistical analysis designed to reflect the sampling scheme, data from a probability sample can be used to extrapolate from the sample to the entire population. This makes probability sampling ideal for descriptive studies, of which the survey is a classic example. Statistical methods that account for the sampling mechanism of a probability sample are said to be *design-based*.

Suppose the finite population has N individuals. Let us denote by x_i the measurement of interest on individual i, for $i = 1, 2, \ldots, N$, and represent the population as the set of measurements $U = \{x_1, x_2, \ldots, x_N\}$. We note that each x_i is a constant, not a random quantity. The *population mean* and *population variance* are defined to be

$$\overline{x}_U = \frac{1}{N} \sum_{i=1}^{N} x_i$$

and

$$s_U^2 = \frac{1}{N-1} \sum_{i=1}^{N} (x_i - \overline{x}_U)^2 ,$$

respectively. Constants such as $\overline{x}_U$ and s_U^2 that are expressed in terms of the entire population are called *population parameters*.

We consider three fundamental types of probability sampling—simple random sampling, stratified random sampling and cluster sampling. The number of individuals in a sample is called the *sample size* and is denoted by n.

Simple Random Sampling

In *simple random sampling*, subjects are chosen in such a way that every sample of a given size has the same nonzero probability of being selected. This method of sampling is most easily implemented when there is a complete roster of the population. For example, suppose the population is all students attending public school in a given school district. It is likely that school authorities would be able to provide a list of students within their jurisdiction. Selecting a simple random sample could be operationalized as follows: number the students on the roster from 1 to N, use a random number table or a computer-based random number generator to produce a random sequence of n numbers between 1 and N inclusive, and then select the corresponding individuals from the list to be in the sample.

A random quantity used to estimate a population parameter is said to be an *estimator* of the parameter. If the *expected value* of an estimator equals the parameter, we say that the estimator is *unbiased*, otherwise *biased*. Let $x_1, x_2, \ldots, x_n$ be a simple random sample from U, where, for convenience of notation, we assume that the sample is comprised of the first n individuals in the population. Under simple random sampling, unbiased estimators of $\overline{x}_U$ and s_U^2 are given by

$$\overline{x}_{\text{srs}} = \frac{1}{n}\sum_{i=1}^{n} x_i \tag{2.3}$$

and

$$s_{\text{srs}}^2 = \frac{1}{n-1}\sum_{i=1}^{n}(x_i - \overline{x}_{\text{srs}})^2, \tag{2.4}$$

which are called the *sample mean* and *sample variance*, respectively. Although $x_1, x_2, \ldots, x_N$ are constants, $\overline{x}_{\text{srs}}$ and s_{srs}^2 are random quantities because they are obtained through probability sampling.

The variance of $\overline{x}_{\text{srs}}$ is

$$\text{var}(\overline{x}_{\text{srs}}) = \left(\frac{N-n}{N}\right)\frac{s_U^2}{n},$$

and an unbiased estimator of $\text{var}(\overline{x}_{\text{srs}})$ is given by

$$\widehat{\text{var}}(\overline{x}_{\text{srs}}) = \left(\frac{N-n}{N}\right)\frac{s_{\text{srs}}^2}{n}. \tag{2.5}$$

The term $(N - n)/N$ is called a *finite population correction factor*. If we take a "sample" of size N, that is, conduct a census, then $n = N$, and so $\overline{x}_{\text{srs}} = \overline{x}_U$ and $\text{var}(\overline{x}_{\text{srs}}) = 0$, which says that the population mean is known precisely.

The estimator $\overline{x}_{\text{srs}}$ is intuitively reasonable, but we can give it a more formal rationale as a weighted average. For the moment, let us set aside the specific case of simple random sampling. In a probability sample, each individual in the population has a known nonzero probability of being selected. For example, everyone may have the same selection probability, or certain subgroups of the population may be targeted for over- or underrepresentation. The *design weight* for each individual in a probability sample is defined to be the reciprocal of the probability of that person being selected. The idea is that the design weight is proportional to the number of people in the population "represented" by the selected individual: the smaller the probability of someone being selected, the larger the number of people that person represents if selected for the sample.

Since each individual in a simple random sample has a selection probability of n/N, the design weight for individual i is $w_i = N/n$, for $i = 1, \ldots, n$. It can be shown that

$$\sum_{i=1}^{n} w_i = N$$

and

$$\sum_{i=1}^{n} \left(\frac{w_i}{\sum_{i=1}^{n} w_i} \right) x_i = \overline{x}_{\text{srs}} .$$

Simple random sampling is intuitively appealing and relatively easy to implement, but it has two significant drawbacks. To illustrate, let us return to the example of public school students and assume that the finite population is a large county. Suppose we are interested in estimating the population mean not only at the county level but also for geographic regions making up the county. With simple random sampling, it could happen by chance that some regions are absent from the sample, making it impossible to estimate the corresponding population means. Another potential problem is that students selected for the sample might be thinly distributed over a large geographic area, resulting in prohibitive costs associated with conducting in-person interviews.

We next consider alternatives to simple random sampling that are designed to address these shortcomings.

Stratified Random Sampling

In *stratified random sampling*, the population is partitioned into groups of individuals, referred to as *strata*, and a simple random sample is selected from each *stratum*. Continuing with the above example of school children, a stratified random sample could be obtained by dividing the county into geographic regions and then selecting a simple random sample of students from each region. This guarantees that all regions are represented, but it does not address the possibility that students selected for the sample might be widely dispersed within a region.

Suppose the population $U = \{x_1, x_2, \ldots, x_N\}$ is divided into m strata that we index by $h = 1, 2, \ldots, m$. Denote by N_h the number of individuals in stratum h, so that $N = \sum_{h=1}^{m} N_h$. Let x_{hi} be the measurement on individual i in stratum h, for $h = 1, \ldots, m$ and $i = 1, \ldots, N_h$. The set of measurements in stratum h is then $U_h = \{x_{h1}, x_{h2}, \ldots, x_{hN_h}\}$. The population mean for stratum h is

$$\overline{x}_{U,h} = \frac{1}{N_h} \sum_{i=1}^{N_h} x_{hi} \, ,$$

and it can be shown that

$$\overline{x}_U = \sum_{h=1}^{m} \left(\frac{N_h}{N} \right) \overline{x}_{U,h} \, .$$

Thus, the overall population mean is a weighted average of stratum-specific population means.

Consider a stratified random sample of the population, where a simple random sample of size n_h is selected from stratum h, so that $n = \sum_{h=1}^{m} n_h$. For simplicity of notation, let us assume that the simple random sample from stratum h is $x_{h1}, x_{h2}, \ldots, x_{hn_h}$. Adding a subscript h to the expression in (2.3), an unbiased estimator of $\overline{x}_{U,h}$ is given by

$$\overline{x}_{\mathrm{srs},h} = \frac{1}{n_h} \sum_{i=1}^{n_h} x_{hi} \, . \tag{2.6}$$

Under stratified random sampling, an unbiased estimator of $\overline{x}_U$ is given by

$$\overline{x}_{\mathrm{str}} = \sum_{h=1}^{m} \left(\frac{N_h}{N} \right) \overline{x}_{\mathrm{srs},h} \, . \tag{2.7}$$

Cluster Sampling

In *cluster sampling*, the population is first partitioned into convenient groupings of individuals called *clusters*, next a probability sample of clusters is selected, and then either a probability sample or a census of each of the selected clusters is obtained. With clustering, respondents tend to be less dispersed geographically than in either simple random sampling or stratified random sampling. Continuing with the above example of school children, we could define all classrooms in the county to be the clusters, select a simple random sample of classrooms, and then interview all students in each of the selected classrooms.

Stratification and clustering can be combined to obtain a *stratified cluster sample*. For instance, we could stratify the county into geographic regions and then select a cluster sample of classrooms within each region. Geographic dispersion can be further reduced by clustering at two or more stages, giving what is called a *multistage cluster sample*. In that setting, the clusters making up the entire population are called *primary sampling units*, each selected primary sampling unit is divided into clusters called *secondary sampling units*, each selected secondary sampling unit is divided into clusters called *tertiary sampling units*, and so on. For example, a two-stage cluster sample of students could be obtained by selecting a simple random sample of all schools in the county (primary sampling units), and then selecting a simple random sample of all classrooms in each of the selected schools (secondary sampling units). Multistage cluster sampling can be combined with stratification to give a *stratified multistage cluster sample*.

A first impression is that the National Health Survey discussed in Section 1.7 has a stratified three-stage design, with stratification according to four geographic regions, and with clustering at the first stage by city, at the second stage by groupings of city blocks, and at the third stage by households. However, probability sampling was not used to select the cities. They were handpicked in a deliberate manner so as to have the survey sample match the sociodemographic makeup of the urban population of the United States, as reflected in the 1930 national census.

A feature of cluster sampling is that individuals in a cluster are likely to have certain characteristics in common. For example, when clustering is based on neighborhood of residence, respondents are likely to have similar sociodemographic profiles. As a consequence, the data from a cluster are often positively correlated, and so the amount of "information" available from a cluster sample is typically less than what would be obtained from a simple random sample or stratified random sample of the same size.

Random Sampling

Probability models are determined, in part, by *model parameters*. For example, a normal distribution is completely characterized by its mean and variance. Such parameters are unknown constants that must be estimated from study data. A random variable that is used to estimate a model parameter is called an *estimator* of the parameter. If the *expected value* of an estimator equals the parameter, we say that the estimator is *unbiased*, otherwise *biased*.

Let X be a random variable and denote its mean and variance by μ and σ^2, respectively. For a given positive integer n, the random variables $X_1, X_2, \ldots, X_n$ are said to be a *random sample* from the probability distribution of X if: (1) each of $X_1, X_2, \ldots, X_n$ has the probability distribution of X, and (2) $X_1, X_2, \ldots, X_n$ are independent random variables, that is, they are mutually uncorrelated.[7] We refer to n as the *sample size* of the random sample. In random sampling, there is no notion of an underlying finite population from which the sample is selected. That role is subsumed by the probability distribution.

The terminology "random sample", although well-established in the literature, is potentially confusing because simple random samples and stratified random samples are not random samples in the sense just defined. To make matters worse, some authors refer to probability samples as random samples.

Suppose $X_1, X_2, \ldots, X_n$ is a random sample from the probability distribution of X. Unbiased estimators of μ and σ^2 are given by

$$\overline{X} = \frac{1}{n}\sum_{i=1}^{n} X_i \tag{2.8}$$

and

$$S^2 = \frac{1}{n-1}\sum_{i=1}^{n}(X_i - \overline{X})^2, \tag{2.9}$$

which are called the *sample mean* and *sample variance*, respectively. The variance of $\overline{X}$ is

$$\mathrm{var}(\overline{X}) = \frac{\sigma^2}{n},$$

and an unbiased estimator of $\mathrm{var}(\overline{X})$ is given by

$$\widehat{\mathrm{var}}(\overline{X}) = \frac{S^2}{n}. \tag{2.10}$$

Statistical methods based on probability models are said to be *model-based*. This includes simple descriptive statistics, but also more complicated approaches such as various types of regression analysis.

Observe that (2.3) and (2.8) are identical as formulas for performing computations, and likewise for (2.4) and (2.9). This means that regardless of whether data for a given study have been collected by simple random sampling or random sampling, the sample means are the same, and likewise for the sample variances. Furthermore, for fixed n, as N increases without bound, the limiting value of (2.5) is (2.10). In light of these observations, we can think of a random sample from a probability distribution as a type of simple random sample from an "infinite" population.

The random variable X is said to have a *Bernoulli distribution* with parameter R if it takes values 0, 1, with $R = \Pr(X = 1)$. The *binomial distribution*, which is a mainstay of the statistical analysis of discrete data, is closely related to the Bernoulli distribution. Specifically, it can be shown that if $X_1, X_2, \ldots, X_n$ is a random sample from the Bernoulli distribution with parameter R, then the random variable $\sum_{i=1}^{n} X_i$ has the binomial distribution with parameters n and R.

Nonprobability Sampling

Nonprobability sampling occupies the middle ground between probability sampling and random sampling. It provides a mechanism for selecting a sample from a finite population without the complexity and cost of probability sampling, while at the same time accommodating a model-based analysis, as would be employed with random sampling. Surveys based on nonprobability sampling are well-established in marketing research and political polling, and are being seen more often in epidemiology, particularly in the form of internet-based surveys. There are several types of nonprobability sampling. Here we focus on two that are of particular relevance to epidemiologic research—convenience sampling and quota sampling.

In *convenience sampling*, individuals are selected from the population because they literally are convenient to recruit. For a survey of school children, the investigators could simply interview a number of interested students attending a school that is easily accessible. Convenience sampling poses some obvious problems of generalizability of findings to the population as a whole. In *quota sampling*, individuals are selected from the population in a deliberate manner, attempting to ensure that the overall sample has a composition that reflects particular aspects of the finite population of interest. Quota sampling was used in parts of the National Health Survey.

In the absence of a probability sampling scheme, there is no possibility of a design-based analysis of data as there would be for a probability sample. However, a probability model can be introduced into the setting of a nonprobability sample to make a model-based analysis possible. This is accomplished by viewing the finite population $\{x_1, x_2, \ldots, x_N\}$ from which the nonprobability sample is selected as a realization of random variables $\{X_1, X_2, \ldots, X_N\}$. The joint probability distribution of $X_1, X_2, \ldots, X_N$ is sometimes referred to as a *superpopulation*.[8] In the simplest case, the random variables are assumed to be a random sample from some probability distribution.

Sampling in Epidemiologic Studies

The method of sampling used in an epidemiologic study depends on whether it is descriptive or etiologic (see Fig. 2.2).

Descriptive studies. Most surveys in epidemiology are conducted using traditional probability sampling of a finite population, followed by a design-based data analysis. Because studies based on vital statistics typically use databases that cover an entire finite population, the subjects in such studies are a census rather than some type of probability or nonprobability sample. From the finite population perspective, this means that population parameters are known precisely. However, if the databases are considered to have a random quality, such as might arise from the way data are collected and recorded, this randomness can be taken into account by conducting a model-based analysis.

Etiologic studies. The samples used in etiologic studies are sometimes difficult to characterize. Suppose, for example, that we are interested in studying patients admitted to hospital following a stroke. One approach to sampling would be to enlist the cooperation of several hospitals in a specific geographic region to help recruit stroke patients for the study. The group of study patients could be viewed as a convenience sample from the finite population of all stroke patients hospitalized in that region over a given time period. Alternatively, under the assumption that the features of interest in "stroke patients" are characterized by a particular probability distribution, the group of study patients could be regarded as a random sample from that probability distribution. Samples for etiologic studies can often be given this dual interpretation.

Regardless of whether the subjects for an etiologic study are viewed as a nonprobability sample or a random sample, the data analysis is ultimately model-based.

It must be emphasized that an etiologic study is more than just an exercise in data collection and statistical analysis. There are larger issues to consider having to do with scientific inference and decision-making, and this requires a clearly-articulated causal framework within which the study is designed, the data generated, and the findings interpreted.[9(p.605–624)]

2.6 Source Populations and Target Populations

Let us refer to the population from which the sample for an epidemiologic study is selected as the *source population* for the study. The population to which the findings from the source population are to be generalized is called the *target population* for the study. In a descriptive study, the source population and the target population are often the same. For example, the source population for the National Health Survey was the population of the United States, and clearly that was also the target population. In an etiologic study, the target population is almost always some larger entity that encompasses the source population. For example, the source population for the Snow cholera study was a specific geographic area of south London. Presumably, the target population Snow had in mind was at least the city of London, perhaps all of England, and maybe even individuals everywhere at risk of cholera from drinking water contaminated with sewage.

2.7 Study Error

There are two types of error in epidemiologic studies—random error and systematic error.

Random Error

The defining characteristic of *random error* is that it is due to chance and therefore unpredictable. Suppose that a study designed to test a certain hypothesis is conducted on two occasions using identical methods. It is possible for one replicate to lead to a correct inference, and for the other to yield to an incorrect inference as a result of random error. For example, consider a study that involves tossing a coin 100 times, where the aim is to test the hypothesis that the coin is fair, that is, has an equal chance of landing heads or tails. Suppose that unknown to the investigator the coin is indeed fair. Imagine that in the first replicate there are 49 heads and 51 tails, which would almost certainly lead to the correct inference that the coin is fair. Now suppose that in the second replicate there are 99 heads and 1 tails, leading to the incorrect inference that the coin is unfair. The erroneous inference made in the second replicate is due to random error, which occurred despite the fact that precisely the same study methods were used for both replicates. Based on the binomial model, the probability of observing the data in the second replicate is 7.97×10^{-29}, a vanishing small number. Although unlikely, these data might arise, and so an incorrect inference is always possible no matter how carefully the study is conducted.

The findings from an epidemiologic study are often presented in terms of a parameter estimate. The random error associated with a parameter estimate can be quantified using its variance. In the epidemiologic literature, a parameter estimate is said to be *precise* when its variance is "small".

Systematic Error

The cardinal feature of *systematic error*, and the characteristic that distinguishes it from random error, is that it is reproducible and therefore predictable. For the most part, systematic error occurs as a result of shortcomings in study methodology. If such problems are not corrected and identical methods are used to replicate the study, the same systematic errors will occur. For example, suppose that in a study of diabetes the laboratory apparatus measuring serum glucose consistently produces values that are larger than the true value. Systematic error in the form of erroneous measurements will be present until the apparatus is repaired. As can be imagined, there is an almost endless array of possibilities for systematic error in epidemiologic studies. Certain study designs are by their very nature more prone to particular types of systematic error than others, as will be discussed in subsequent chapters. With careful attention to study methods it is possible to minimize systematic error, at least those sources of systematic error that become known to the investigators.

When systematic error is present, parameter estimates based on the epidemiologic study will usually be biased, in the statistical sense of the term. Epidemiology has borrowed the term *bias* from the statistical literature, using it as a synonym for systematic error. Accordingly, when an epidemiologic study is subject to systematic error we say that the parameter estimate in question is biased or, rather more loosely, that the study is biased. A parameter estimate is said to be *valid* when it is unbiased.

Breaches of validity are usually classified as being one of three types, namely, selection bias, information bias and confounding. *Selection bias* is due to the manner in which subjects are selected for a study, as well as the factors that relate to their ongoing participation. *Information bias* arises from the way in which information is collected on study participants and subsequently handled. When information bias impacts a continuous variable, the effect is called *measurement error*, and when the variable is discrete, the effect is referred to as *misclassification bias* or simply *misclassification*.

Confounding (or *confounding bias*) arises due to the presence of associations among variables that have the potential to distort the measurement of causal associations of interest. For example, in the Snow cholera study we need to at least consider the possibility that customers of the Southwark and Vauxhall Company were (somehow) more susceptible than customers of the Lambeth Company to the lethal consequences of cholera infection. For instance, perhaps clients of the Southwark and Vauxhall Company were older and more frail compared to clients of the Lambeth Company and therefore more likely to succumb to cholera. If so, part of the increased mortality rate among customers of the Southwark and Vauxhall Company compared to customers of the Lambeth Company could be accounted for by their differing age distributions. In that case, we would say that "age" is a *confounder* of the association between "water company" and "cholera mortality", and steps would need to be taken to "control for" the confounding effects of age in the data analysis. We will have much more to say about confounding in subsequent chapters.

References

1. Black WC, Welch HG. Screening for disease. *Am J Roentgenol* 1997;**168**(1):3–11.
2. Cole P, Morrison AS. Basic issues in population screening for cancer. *J Natl Cancer Inst* 1980;**64**(5):1263–72.
3. Herman CR, Gill HK, Eng J, et al. Screening for preclinical disease: test and disease characteristics. *Am J Roentgenol* 2002;**179**(4):825–31.
4. Hill AB. The environment and disease: association or causation? *Proc R Soc Med* 1965;**58**(5):295–300.
5. Hernán MA. The C-word: scientific euphemisms do not improve causal inference from observational data. *Am J Public Health* 2018;**108**(5):616–9.
6. Lohr SL. *Sampling: design and analysis*. 2nd ed. Boston: Brooks/Cole; 2010.
7. Hogg RV, McKean JW, Craig AT. *Introduction to mathematical statistics*. 8th ed. Boston: Pearson; 2019.
8. Kalsbeek W, Heiss G. Building bridges between populations and samples in epidemiological studies. *Annu Rev Public Health* 2000;**21**:147–69.
9. Greenland S. The causal foundations of applied probability and statistics. In: Geffner H, Dechter R, Halpern JY, editors. *Probabilistic and causal inference: the works of Judea Pearl*. New York: Association for Computing Machinery; 2022.

CHAPTER

Prevalence Studies

3

3.1 Prevalence

Consider a geographically-defined population, such as a city or a country, at a specific time t. Anyone in the population who is a case of a given disease at that time is said to be a *prevalent case* of the disease at time t. Certain diseases are chronic, meaning that once they develop, there is no cure and symptoms may persist indefinitely. All members of the population at time t who are cases of such a disease would be counted as prevalent cases. On the other hand, there are diseases that afflict individuals acutely and then resolve more or less completely. If someone who experienced one of these diseases has recovered by time t, that person would not be considered a prevalent case.

Let C be the number of prevalent cases and let N be the number of individuals in the population at time t. The *point prevalence* of the disease at time t is defined to be the proportion of the population who are cases of the disease at that time, that is,

$$P = \frac{C}{N}. \tag{3.1}$$

The point prevalence is a measure of disease burden in the population and plays an important role in public health planning. In the notation of Section 2.5, the population can be represented by $\{x_1, x_2, \ldots, x_N\}$, where $x_i = 1$ if individual i is a case at time t, and $x_i = 0$ if not, for $i = 1, \ldots, N$. Then $C = \sum_{i=1}^{N} x_i$ and, by definition, P is the population mean.

The *point prevalence odds* of the disease at time t is defined to be

$$PO = \frac{P}{1-P} = \frac{C}{N-C}. \tag{3.2}$$

It follows from the identity

$$P = \frac{PO}{1+PO}$$

that P and PO have equivalent mathematical content. When P is "small", we have the approximation

Epidemiologic Methods. https://doi.org/10.1016/B978-0-44-318780-3.00009-9
Copyright © 2023 Elsevier Inc. All rights reserved.

$$PO \approx P\,. \tag{3.3}$$

Although appearing somewhat out of place in a discussion about disease in a population, odds terminology is well established in games of chance, an observation we return to in Section 4.1.

Provided the population is sufficiently small, it might be feasible to determine the point prevalence by collecting data on all members of the population, that is, by conducting a census. In practice, the point prevalence is usually estimated by carrying out a type of survey called a *prevalence study*. This involves selecting a probability sample from the population, collecting data on each individual in the sample regarding case status, and performing a design-based data analysis to arrive at an estimate of the point prevalence. Determining the case status of respondents might rely on their recall of previously-established diagnoses or it could involve the collection of diagnostic data expressly for the prevalence study, possibly including physical measurements and laboratory tests. As we saw in Section 2.5, the attraction of a probability sample is that it is structured to be representative of the population, which fits perfectly with the descriptive nature of a prevalence study.

A prevalence study is often said to be a "snapshot" of the population, the implication being that sample selection and data collection take place at a single time point. In practice, these activities typically stretch over a number of weeks or months. This means that for a prevalence study to be interpreted as a snapshot, it must be assumed that the population is more or less unchanging over the time period during which the prevalence study is conducted. This assumption is usually implicit in published reports of prevalence studies but is rarely discussed.

It is intuitively clear that the point prevalence of a disease is determined, at least in part, by the rapidity with which new cases appear in the population and the typical duration of their illness. The faster new cases emerge and the longer the disease persists, the greater the point prevalence. This relationship is demonstrated more rigorously in Section 5.5. Imagine a disease that is uniformly lethal but can be either slowly or rapidly progressive. Individuals developing the rapidly progressive form exit the population relatively quickly compared to those with slowly progressive disease. As a consequence, rapidly progressive cases will be underrepresented in the population at any given time compared to slowly progressive cases. This is not a shortcoming of the point prevalence as a measure of disease burden, but rather a feature to be kept in mind when interpreting the findings from a prevalence study.

As an alternative to the point prevalence, let us ask instead what proportion of the population at time t were cases at any point during a given time period ending at time t. This is referred to as the *period prevalence* of the disease at time t. Typical time periods are the preceding month and past year, which yield the *1-month prevalence* and *1-year prevalence*, respectively. It is also meaningful to ask whether a member of the population at time t was ever a case, which leads to the *lifetime prevalence* of the disease at time t. Evidently, if someone was a case in the past month, that person was a case in the past year. This means that the 1-month prevalence is less than or equal to the 1-year prevalence, which in turn is less than or equal to the lifetime prevalence. For certain diseases, a period prevalence may be more relevant to public health planning than the point prevalence. This is particularly so for diseases that have a relapsing-remitting course, such as certain mental disorders. The National Health Survey discussed in Section 1.7 reported on the 1-year prevalence of disability.

Example: National Comorbidity Survey Replication

The National Comorbidity Survey Replication was a prevalence study of mental disorders in the United States conducted during 2001–2003.[1,2] Subjects were a representative sample of the English-speaking, noninstitutionalized, civilian population 18+ years old. The prevalence study had a stratified four-stage design, with stratification by geographic region of the country. For stage one, primary sampling units consisting of one or more counties were constructed; a probability sample of 62 of them was selected. For stage two, each selected primary sampling unit was divided into secondary sampling units consisting of 50 to 100 households; a probability sample of 1,001 of them was selected. For stage three, all households in each selected secondary sampling unit were rostered and a probability sample of households was selected. For stage four, eligible occupants of each selected household were listed, and primary and secondary respondents were chosen using a selection grid. Each respondent was administered a highly structured, computer-assisted, in-person questionnaire by a trained nonclinician interviewer. The process lasted between 90 minutes and six hours depending on the history of mental disorders. A total of 9,282 interviews were completed. The 1-year prevalence of major depressive disorder was 6.6%. ◊

3.2 Bias in Prevalence Studies

Selection Bias

Sampling bias arises when the probability sampling procedures used to select subjects are flawed, resulting in a nonrepresentative sample of the population. For example, if households are contacted during daytime hours and those who tend to be at home during the day have a different point prevalence of disease than who are usually away during the day, the sample will not be representative of the population.

Nonparticipation bias (or *nonresponse bias*) occurs when individuals belonging to certain subgroups of the population refuse to participate in a prevalence study, leading to a nonrepresentative sample of the population. For example, if the point prevalence increases with age and younger members of the population are more likely to refuse participation than those who are older, a simple random sample will tend to overestimate the point prevalence. Provided at least some information can be collected on those refusing to be interviewed, it may be possible to account for nonparticipation as part of a design-based analysis by modifying design weights accordingly.

Information Bias

Interviewer bias occurs when the person collecting information from a respondent elicits responses in a way that results in case status being assigned incorrectly. For example, an interviewer might consistently overemphasize certain items on a questionnaire due to curiosity about particular topics. This problem can be mitigated by using highly-structured questionnaires that permit little deviation from a standardized protocol.

Response bias arises when respondents provide incorrect information, either intentionally or unintentionally. Eliciting information from respondents in a manner that is nonthreatening and supportive can help to minimize this problem.

Misclassification of Case Status

Misclassification can affect any of the discrete variables in a prevalence study. Here we focus on misclassification of case status. At any given moment, each individual in the population is either a case or not. Let us denote case status by the variable D, with $D = 1$ indicating that someone is a case, and $D = 0$ a noncase. We anticipate the discussion of screening programs in Chapter 15 and use the term "test" to refer to the overall mechanism (interviewer, questionnaire, diagnostic algorithm, etc.) involved in assigning case status. A test yields one of two outcomes—*positive* for someone presumed to be a case, and *negative* for someone presumed to be a noncase. Let us denote the outcome of the test by the variable $D^\star$, with $D^\star = 1$ indicating a positive test, and $D^\star = 0$ a negative test. An infallible test is one that categorizes respondents perfectly with respect to case status, that is, behaves like a diagnostic assessment. Because tests are usually fallible, there will inevitably be cases with a negative test, and noncases with a positive test, resulting in misclassification of case status. Table 3.1 gives the counts for an entire population according to case status and test outcomes, as well as the designations used to refer to test categories. For instance, when a case has a negative test, the test is said to be *false negative*, and when a noncase has a positive test, we say the test is *false positive*.

In the notation of Table 3.1, the number of prevalent cases is $C = A_1 + A_0$ and the number of individuals with a positive test is $A_1 + B_1$. The point prevalence is

$$P = \frac{A_1 + A_0}{N},$$

and the proportion of the population with a positive test is

$$P^\star = \frac{A_1 + B_1}{N}. \tag{3.4}$$

Table 3.1 Population counts of prevalence.

Test	$D = 1$	$D = 0$	Total
$D^\star = 1$	A_1 (true positive)	B_1 (false positive)	$A_1 + B_1$
$D^\star = 0$	A_0 (false negative)	B_0 (true negative)	$A_0 + B_0$
Total	$A_1 + A_0$	$B_1 + B_0$	N

A prevalence study yields data corresponding to the far-right column of Table 3.1, allowing us to estimate $P^\star$ but not P. Let us adopt the convention of adding a star to notation to denote quantities subject to misclassification. We now think of $D^\star$ and $P^\star$ as the misclassified counterparts of D and P.

The question arises as to how accurately the test reflects case status. This property of a test is called its *validity*. Ideally, we would like there to be no false negative tests and no false positive tests, that is, $A_0 = 0$ and $B_1 = 0$. In that situation, the test is effectively a diagnostic assessment and $P^\star = P$. For a dichotomous test, the most widely-used measures of validity in epidemiology are sensitivity and specificity. The *sensitivity* of a test, denoted here by S, is the proportion of cases in the population with a positive test; and the *specificity* of a test, denoted here by T, is the proportion of noncases in the population with a negative test:

$$S = \frac{A_1}{A_1 + A_0}$$

$$T = \frac{B_0}{B_1 + B_0}.$$

When there are no false negative tests, the sensitivity is 100%; and similarly, when there are no false positive tests, the specificity is 100%.

Although a prevalence study does not provide the data needed to estimate the sensitivity and specificity of a test, a *validity study* of the test can be conducted separately. This involves recruiting samples of known cases and noncases, administering the test to each of them, and then estimating the sensitivity and specificity. For example, the cases and noncases could come from a health care facility that treats individuals with the disease of interest. The sensitivity of the test might depend on the stage of the disease and other factors, and so it is important that the case sample be representative of all prevalent cases in the population.

In practice, there is a trade-off between the sensitivity and specificity of a test. This is most easily understood by examining the situation where the test outcome is a continuous variable, which we term the *test score*. Fig. 3.1 shows two bell-shaped curves, one depicting the distribution of test scores for cases, and the other the distribution for noncases. A cut-off, shown as a vertical line, needs to be chosen such that individuals with a test score above the cut-off are said to have a positive test, and those with a score below the cut-off a negative test. Because the curves overlap, some cases will have a negative test (false negative) and some noncases a positive test (false positive). The sensitivity (specificity) of the test can be increased by lowering (raising) the cut-off, but this comes at the expense of a decrease in specificity (sensitivity).

Example: National Comorbidity Survey Replication

A validity study was performed in conjunction with the National Comorbidity Survey Replication (NCS-R) discussed in Section 3.1 in which a probability sample of 335 respondents underwent reassessment by a clinician to determine the 1-year preva-

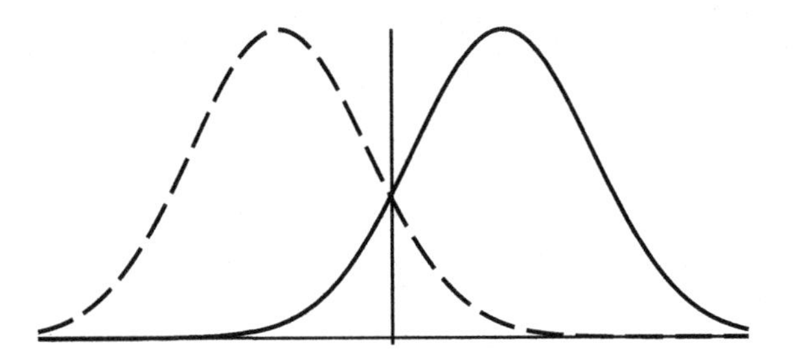

FIGURE 3.1

Distribution of test scores for cases and noncases (cases: *solid*; noncases: *dash*; cut-off: *vertical*).

Table 3.2 Misclassified data for hypothetical prevalence study: sensitivity 80% and specificity 90%.

Test	$D = 1$	$D = 0$	Total
$D^\star = 1$	80 = 80% × 100	90	170
$D^\star = 0$	20	810 = 90% × 900	830
Total	100 = 10% × 1,000	900	1,000

lence of selected mental disorders.[2] With clinician diagnosis as the gold standard, the sensitivity and specificity of the computer-based diagnostic algorithms used in the NCS-R to estimate the 1-year prevalence of major depressive episode were 55% and 95%, respectively. ◊

Example: Point Prevalence in Hypothetical Population

Consider a hypothetical population of $N = 1,000$ individuals, where the point prevalence of a certain disease is $P = 10\%$. It follows that there are 100 $(= 10\% \times 1,000)$ cases and 900 noncases in the population. Suppose everyone in the population is administered a test that has a sensitivity of $S = 80\%$ and specificity of $T = 90\%$. Of the 100 cases, 80 $(= 80\% \times 100)$ will have a true positive test and 20 a false negative test. Of the 900 noncases, 810 $(= 90\% \times 900)$ will have a true negative test and 90 a false positive test. As shown in Table 3.2, this gives 170 individuals with a positive test and 830 with a negative test. The misclassified point prevalence is

$$P^\star = \frac{170}{1,000} = 17\%,$$

which overestimates the true point prevalence by a factor of $P^\star/P = 1.7$. ◊

Returning to the general discussion, it is shown in Appendix B that for a prevalence study where the sensitivity and specificity of the test are S and T, respectively, the misclassified point prevalence is

$$P^\star = 1 - T + (S + T - 1)P. \qquad (3.5)$$

Fig. 3.2 shows the ratio $P^\star/P$ as a function of P, for selected values of sensitivity and specificity. For values of P likely to be seen in prevalence studies, $P^\star$ will overestimate P, and perhaps to a substantial degree.

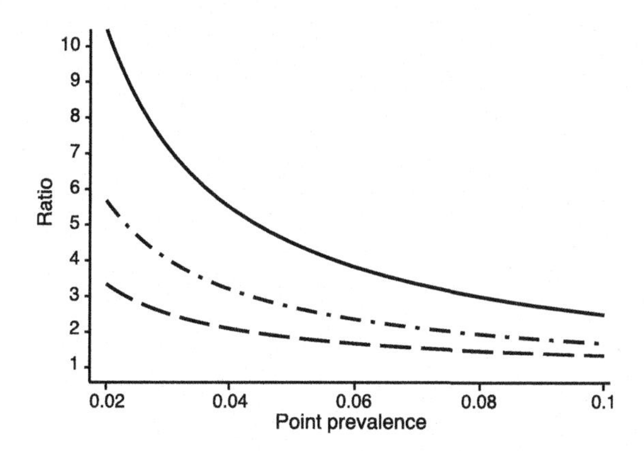

FIGURE 3.2

$P^\star/P$ as a function of P (sensitivity 70% and specificity 80%: *solid*; sensitivity 80% and specificity 90%: *dashdot*; sensitivity 90% and specificity 95%: *dash*).

References

1. Kessler RC, Berglund P, Chiu WT, et al. The U.S. National Comorbidity Survey Replication (NCS-R): design and field procedures. *Int J Methods Psychiatr Res* 2004;**13**(2):69–92.
2. Kessler RC, Berglund P, Demler O, et al. The epidemiology of major depressive disorder: results from the National Comorbidity Survey Replication (NCS-R). *JAMA* 2003;**289**(23):3095–105.

CHAPTER

Cohort Studies

4

In a *cohort study*, a group of individuals called a *cohort* is followed over time for the occurrence of an *outcome* of interest. Members of a cohort are chosen because they have certain attributes in common, such as demographic characteristics, residence in a geographic location, health status, or occupational history. For example, the cohort in the Snow cholera study consisted of all residents of a certain area of London, and the cohort for the MRC streptomycin trial was comprised of tuberculosis patients satisfying specific clinical criteria. The outcomes that might be of interest in a cohort study are virtually unlimited and include, for example, developing a disease, recovering from a disease, dying of a disease, being admitted to hospital, being discharged from hospital, responding to therapy, and so on. For concreteness, throughout the remainder of this chapter—indeed throughout the rest of the book—we focus on developing a disease as the outcome of interest. In light of our discussion of the natural history of disease in Section 2.1, the expression "developing a disease" is problematic. Let us instead resort to the admittedly awkward phrase "becoming a case" as an alternative. Only those individuals who are at risk of becoming a case are eligible to be included in the cohort. This immediately excludes prevalent cases, but also eliminates any noncases who cannot become a case. As an example of the latter, a woman who has had a hysterectomy would not be included in a cohort study where the outcome is endometrial cancer.

As can be seen from Table 2.1, of the eleven studies presented in Chapter 1, three are listed as cohort studies, two as randomized trials, and two as nonrandomized trials. We will see in Chapter 11 that randomized trials and nonrandomized trials are just special types of cohort studies. This means that seven of the eleven studies in Chapter 1 have a cohort design.

There are two types of "time" in a cohort study: calendar time and what we will refer to as personal time. *Calendar time* is the external type of time that applies equally to all individuals. We can think of it as being measured using, for example, the Gregorian calendar. Let us assume for convenience that follow-up of each individual in the cohort begins immediately after recruitment into the study. For some cohort studies, all subjects are recruited at the same calendar time, whereas for others, recruitment takes place over a defined calendar time period, giving what is referred to as *staggered entry* into the cohort. For example, the entire workforce in a manufacturing plant might be considered enrolled in a cohort study on the day a new fabricating process is implemented, while the patients in a consecutive series of admissions to a regional hospital might be recruited into a cohort study over several weeks or months.

Epidemiologic Methods. https://doi.org/10.1016/B978-0-44-318780-3.00010-5
Copyright © 2023 Elsevier Inc. All rights reserved.

Personal time is an internal type of time that applies specifically to each individual in the cohort study. We can think of it as being measured by a clock that begins for each individual at the start of follow-up of that individual. The unit of time adopted for a cohort study is a matter of convenience or convention. In this chapter and throughout the book, calendar time and personal time will be assumed to be calibrated in years unless stated otherwise. In what follows, "time" not otherwise specified will mean calendar time.

The *administrative length of follow-up* of a cohort member is the maximum amount of personal time that the individual can potentially experience in the study. It is determined by the study design and is known before the start of follow-up. The *actual length of follow-up* of a cohort member is the amount of personal time that the person in fact experiences in the study. It is determined by forces outside the study and becomes known only after follow-up of that person has ended. Evidently, the actual length of follow-up is less than or equal to the administrative length of follow-up. For example, if the study design specifies that subjects will be followed for at most five years after recruitment, the administrative length of follow-up is five years. A cohort member who remains under observation for precisely two years has an actual length of follow-up of two years.

By definition, all subjects in a cohort study are noncases at the start of follow-up. A subject who becomes a case during the course of follow-up is said to be an *incident case*. Once someone becomes a case, follow-up of that individual ends, at least as far as the cohort study is concerned. This means that there are no prevalent cases in a cohort study. A subject who does not become a case during follow-up is said to be *censored*. Some subjects are censored simply because they reach their administrative length of follow-up without becoming a case. Others are censored because they cease to be under observation before their administrative length of follow-up is reached, even though they have not become a case. Censored individuals of this type are said to be *lost to follow-up*. There are many reasons for someone being lost to follow-up, such as refusal to continue as a study subject, relocation out of the study area with loss of contact, and death from a cause other than the disease of interest. Being lost to follow-up raises the question of whether the individual concerned might have become a case had follow-up continued as originally planned.

Often the aim of a cohort study is to compare the experience of two or more groups of individuals to determine whether an *exposure* influences becoming a case. The term exposure is used here in a general sense and includes almost any factor that might affect whether someone becomes a case. Examples of exposure are: medical conditions such as elevated blood pressure and a history of cancer; lifestyle habits such as cigarette smoking and alcohol consumption; and occupational factors such as working with dangerous machinery and contact with toxic chemicals. Also considered to be exposures are such sociodemographic characteristics as age, gender, ethnicity and income level.

Cohort studies can be categorized according to the time frame during which data are collected. In a *prospective cohort study*, subjects are recruited and follow-up begins at some point in the near future. This allows the investigators to select subjects appropriate to the aims of the study, gather baseline data, and put a process in place that maximizes the chances of maintaining contact. Cohort studies may require large numbers of subjects, especially when the outcome of interest is a rare occurrence, making it necessary to follow the cohort for an extended period of time so that there are sufficient cases for a meaningful statistical analysis. This means that prospective studies can be very costly and may not provide answers to pressing health issues in a timely fashion.

In a *retrospective cohort study*, an existing longitudinal database is used to assemble a cohort as it would have been constituted in the past. The cohort is then followed "in the database" until a more recent time. This type of cohort study is commonly seen in occupational epidemiology where extensive databases on industrial workforces are often available. The retrospective design offers an alternative to the prospective approach that is economical and yields results relatively promptly. In theory, the findings from a retrospective cohort study are the equivalent of the corresponding prospective cohort study. The reality is that it would be unusual for the database used in a retrospective cohort study to contain all the information that might be desired, making the retrospective approach a practical but generally less informative option than its prospective counterpart.

Another way of categorizing cohort studies is according to whether they are closed or open. This distinction is considered in the next two sections.

4.1 Closed Cohort Studies

A *closed cohort study* is characterized by two features: (1) each subject has the same administrative length of follow-up, and (2) there are no losses to follow-up. This means that each member of the cohort is followed until that individual becomes a case or, failing that, the administrative length of follow-up is reached. Thus, in a closed cohort study, censoring is due exclusively to reaching the administrative length of follow-up without becoming a case.

Fig. 4.1 depicts a closed cohort study with three subjects, labeled A, B and C. The horizontal axis is calibrated in calendar time and, for convenience, has been initialized at zero. Each horizontal line segment represents the experience of a given subject over the course of follow-up. A solid circle indicates that the person became a case, and a hollow circle denotes censoring. Each of the three subjects began follow-up at a different calendar time, and so there was staggered entry. Subject A was censored, whereas subjects B and C became cases. Because subject A did not become a case, the administrative length of follow-up for subject A (hence the common administrative length of follow-up for all subjects) was four years.

Although not labeled as such, personal time is implicit in each of the horizontal line segments. Suppose a medical innovation with the potential to impact the disease became available at calendar time 3.5. This might have affected the outcome of subjects A and C, but not subject B, who had already become a case. This feature would need to be considered in the data analysis and is a reason for distinguishing between

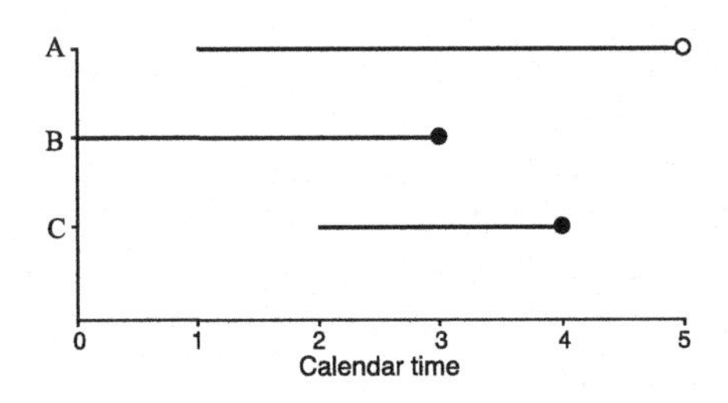

FIGURE 4.1

Closed cohort study with staggered entry (case: *solid circle*; censoring: *hollow circle*).

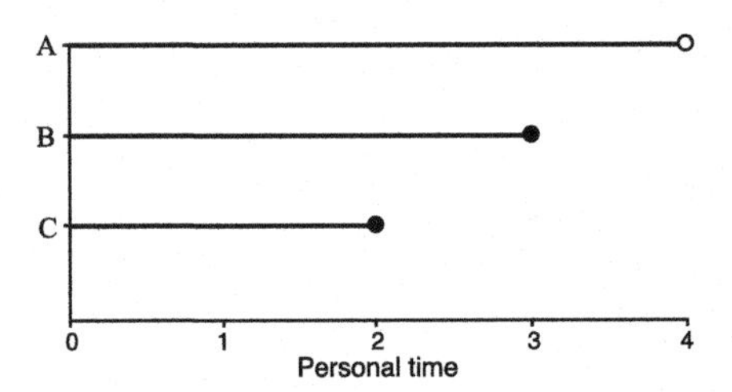

FIGURE 4.2

Closed cohort study without staggered entry obtained by collapsing Fig. 4.1 over calendar time (case: *solid circle*; censoring: *hollow circle*).

calendar time and personal time. On the other hand, if there were no such medical innovation or other external factor that might have had an effect on the disease, staggered entry can be ignored. This allows us to collapse Fig. 4.1 over calendar time to obtain Fig. 4.2, where the horizontal axis is now calibrated in personal time.

Binomial Model

Consider a closed cohort study of n individuals and let us denote by Δ the administrative length of follow-up. Let a be the number of incident cases occurring during follow-up and let $b = n - a$. An obvious choice of probability distribution for analyzing data from such a study is the binomial distribution with parameters n and R_Δ, where R_Δ is the probability of becoming a case. Being a probability, R_Δ is a dimensionless quantity with a value between 0 and 1 inclusive. In epidemiology, R_Δ is referred to as the Δ-*year risk* of becoming a case. Unless there is a possibility of confusion, we usually drop the reference to Δ, writing R in place of R_Δ and referring to R simply as the *risk*.

Recall from Section 2.5 that the binomial distribution with parameters n and R is the probability distribution that results from summing the random variables in a random sample of size n from the Bernoulli distribution with parameter R. The probability function for the binomial distribution is

$$\Pr(a) = \binom{n}{a} R^a (1 - R)^{n-a},$$

for $a = 0, 1, \ldots, n$, where $\binom{n}{a}$ is a binomial coefficient. By definition, random variables in a random sample have the same probability distribution and are mutually uncorrelated. This means that in order for the binomial distribution to be applicable to a closed cohort study, the following two conditions must be satisfied: (1) all subjects have the same risk, and (2) subjects behave independently with respect to becoming a case.

In a typical cohort, it is unrealistic to assume that everyone has the same risk of becoming a case. This assumption can be made more defensible by stratifying the cohort into separate groups in a way that makes the required uniformity more likely. For example, if we know (or can determine from study data) that all males have a similar risk, and likewise for all females, we could analyze male and female cohorts separately. The independence assumption is reasonable in certain settings but not in others. For instance, in a cohort study of an infectious disease in a classroom of school children, it is likely that when one student develops the infection, the other students will be at greater risk.

From the perspective of the binomial model, a is a random variable and n is a known parameter. The random variable $\widehat{R} = a/n$, referred to in epidemiology as the *incidence proportion* (or *cumulative incidence*), is an unbiased estimator of R. For simplicity, we henceforth identify an estimator with its expected value. This allows us to write

$$R = \frac{a}{n}. \tag{4.1}$$

Since a/n is a proportion, and both the numerator and denominator have the units "persons", we see that once again R is a dimensionless quantity with a value between 0 and 1 inclusive.

There is an alternative to R as a way of parameterizing the binomial distribution. The *(incidence) odds* of disease is defined to be

$$W = \frac{R}{1 - R} = \frac{a}{n - a}.$$

We observe that

$$R = \frac{W}{1 + W},$$

which shows that R and W are mathematically equivalent. As remarked in Section 3.1, odds terminology is well established in games of chance. For example, the probability of picking an ace at random from a deck of cards is

$$R = \frac{4}{52} = \frac{1}{13},$$

which can be read as “1 out of 13”. The corresponding odds is

$$\frac{R}{1-R} = \frac{4/52}{1-(4/52)} = \frac{1}{12},$$

which can be read as “1 to 12” and alternatively written as 1:12. An attractive feature of the odds is that it satisfies a reciprocal property: if W is the odds of becoming a case, the odds of not becoming a case is

$$\frac{1-R}{1-(1-R)} = \frac{1}{W}.$$

In the card example, the odds of not picking an ace is

$$\frac{(48/52)}{1-(48/52)} = 12,$$

that is, “12 to 1”.

Measures of Effect

In keeping with earlier remarks, often the aim of a closed cohort study is not only to estimate the risk of becoming a case, but also to examine whether a particular exposure increases or decreases that risk. Except where indicated otherwise, we make the important simplifying assumption that the exposure status of each member of the cohort is fixed at the start of follow-up. In other words, each individual is considered to be either “exposed” or “unexposed”, and this status remains unchanged throughout the study. More general (and more complicated) statistical methods are available that permit exposures to be *time-dependent*.

Consider a closed cohort study in which there are n_1 exposed subjects and n_0 unexposed subjects, so that $n = n_1 + n_0$. We refer to the exposed subjects as the *exposed cohort*, and to the unexposed subjects as the *unexposed cohort*. Suppose that over the course of follow-up there are a_1 incident cases in the exposed cohort and a_0 incident cases in the unexposed cohort, so that $a = a_1 + a_0$. Let $b_1 = n_1 - a_1$ and $b_0 = n_0 - a_0$, so that $b = b_1 + b_0$. We assume that the exposed cohort is governed by the binomial distribution with parameters n_1 and R_1, and similarly that the unexposed cohort is governed by the binomial distribution with parameters n_0 and R_0. Consistent with (4.1), we have

$$R_1 = \frac{a_1}{n_1}$$
$$R_0 = \frac{a_0}{n_0}. \tag{4.2}$$

Let us denote exposure status by the variable E, with $E = 1$ indicating that someone is exposed, and $E = 0$ unexposed. The preceding notation is summarized in Table 4.1. Any two of the columns for cases, noncases and persons can be used to derive the third. For typographical convenience, we usually drop the column for non-cases when presenting data from closed cohort studies.

Table 4.1 Data and risks for closed cohort study.

Exposure	Cases	Noncases	Persons	Risk
$E = 1$	a_1	b_1	n_1	R_1
$E = 0$	a_0	b_0	n_0	R_0
Overall	a	b	n	R

There are three basic measures of effect for a closed cohort study. The *risk ratio*, *(incidence) odds ratio* and *risk difference* are defined to be

$$RR = \frac{R_1}{R_0} = \frac{a_1/n_1}{a_0/n_0}$$

$$OR = \frac{R_1/(1-R_1)}{R_0/(1-R_0)} = \frac{a_1/b_1}{a_0/b_0} \tag{4.3}$$

$$RD = R_1 - R_0 = \frac{a_1}{n_1} - \frac{a_0}{n_0},$$

respectively. If we assume that the risk of becoming a case is "small", sometimes referred to as the *rare disease assumption*, OR is approximately equal to RR, written as $OR \approx RR$.

We say that the exposure-disease association is *null* if $R_1 = R_0$, *positive* if $R_1 > R_0$, and *negative* if $R_1 < R_0$. The *null value* of each of the above measures of effect is the value corresponding to $R_1 = R_0$. Thus, RR and OR have the null value 1, and RD the null value 0. It is shown in Appendix C that the exposure-disease association is null if and only if

$$OR = RR = 1 \quad \text{and} \quad RD = 0, \tag{4.4}$$

positive if and only if

$$OR > RR > 1 \quad \text{and} \quad RD > 0, \tag{4.5}$$

and negative if and only if

$$OR < RR < 1 \quad \text{and} \quad RD < 0. \tag{4.6}$$

Observe that when the exposure-disease association is positive (negative), each measure of effect is greater (less) than its null value. Also observe that when there is a nonnull exposure-disease association, the odds ratio is further from the null value than the risk ratio is. When risks are "small", the exposure-disease association can appear to be null when assessed using the risk difference, but not when evaluated using the risk ratio or odds ratio. This is illustrated in the examples below.

Example: SARS Study

Severe acute respiratory syndrome (SARS) is an infectious disease caused by a coronavirus. A closed cohort study of the association between gender and death from SARS was conducted in which 1,755 individuals infected during the SARS epidemic in Hong Kong in 2003 were followed until death or recovery.[1] The mortality risk associated with a potentially fatal illness is traditionally called the *case fatality rate*. Table 4.2 gives the data and case fatality rates for the SARS study, stratified by gender. The rate for males is greater than that for females, and this is reflected in the measures of effect: $RR = 1.66$, $OR = 1.85$ and $RD = 0.0873$. ◊

Table 4.2 Data and case fatality rates for SARS study.

Gender	Deaths	Persons	Rate[a]
Male	170	776	21.9
Female	129	979	13.2
Overall	299	1,755	17.0

[a] *case fatality rate per 100.*
Data from Reference 1.

Example: Down Syndrome Study

Down syndrome (or trisomy 21) is a genetic disorder caused by the presence of an extra copy of chromosome 21. A study of the association between birth order and Down syndrome was conducted using data on live births in Michigan for 1950–1964 obtained from national and state vital statistics databases and institutional records.[2] Birth order of a newborn is defined to be: first offspring of the mother, second offspring of the mother, and so on. Strictly speaking, this study does not have a closed cohort design because newborns with Down syndrome are prevalent cases at the time of birth (not incident cases occurring subsequent to conception), and losses to follow-up after conception due to miscarriages and induced abortions are not taken into account. For purposes of illustration, we ignore these issues and treat the Down syndrome study like a closed cohort study.

The proportion of newborns with a given congenital anomaly (that is, the point prevalence of the congenital anomaly at birth) is traditionally called the *congenital anomaly rate*. Table 4.3 gives the data and congenital anomaly rates for the Down syndrome study, stratified by birth order. The rate for newborns whose birth order is third or greater is larger than that for newborns whose birth order is first or second. This is reflected in the measures of effect: $RR = 1.858$, $OR = 1.859$ and $RD = 0.000536$. Since Down syndrome is a very rare occurrence, the risk ratio and odds ratio are equal for practical purposes, and the risk difference is small enough to be mistaken for the null value. ◊

Table 4.3 Data and congenital anomaly rates for Down syndrome study.

Birth order	Cases	Live births	Rate[a]
Third or greater	1,546	1,330,100	116.2
First or second	877	1,401,629	62.6
Overall	2,423	2,731,729	88.7

[a] *congenital anomaly rate per 100,000.*
Data from Reference 2.

4.2 Open Cohort Studies

The closed cohort design is based on the premise that there are no losses to follow-up. This may be a reasonable expectation when the period of follow-up is short, but over longer time frames it is usually unrealistic. In practice, subjects sometimes end participation prior to reaching their administrative length of follow-up for any of a host of reasons, as remarked above. Under certain circumstances it may be possible to salvage the closed cohort paradigm. If subjects leave the study almost immediately after follow-up begins, we might consider dropping them from the study data. At the other extreme, if subjects remain under observation without becoming cases until just before reaching their administrative length of follow-up, we could decide to treat them as if the administrative length of follow-up had been reached. But what about subjects who exit the study part way through their follow-up without becoming a case? The experience of these individuals tells us something about the risk of becoming a case, but the information is incomplete.

In an *open cohort study*, the two restrictions placed on a closed cohort study are removed: subjects are not required to have the same administrative length of follow-up, and losses to follow-up are permitted. As before, staggered entry is allowed, each subject has an administrative length of follow-up that is determined by the study design, and each individual either becomes a case or is censored. However, in an open cohort study, as opposed to a closed cohort study, censoring may be due to being lost to follow-up as well as reaching the administrative length of follow-up without becoming a case. Fig. 4.3 depicts a cohort study that is identical to the closed cohort presented in Fig. 4.1 except that now subject C is censored. Since the administrative length of follow-up for subject A is at least four years long, and subject C was censored after only two years, Fig. 4.3 represents an open cohort study. For reasons mentioned above in connection with closed cohort studies, it may be appropriate to collapse Fig. 4.3 over calendar time and relabel the horizontal axis as personal time, which gives Fig. 4.4.

For certain diseases, such as influenza or stroke, it is possible to have repeated episodes. If "becoming a case" is defined to be the occurrence of the first episode, follow-up ceases once that outcome occurs. Alternatively, becoming a case could be taken to be the occurrence of a second episode, for example. For such a study, follow-up begins once the first episode is concluded because an individual is at risk

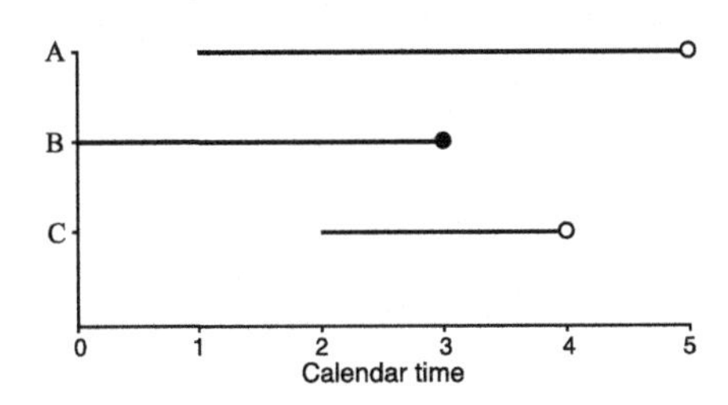

FIGURE 4.3

Open cohort study with staggered entry (case: *solid circle*; censoring: *hollow circle*).

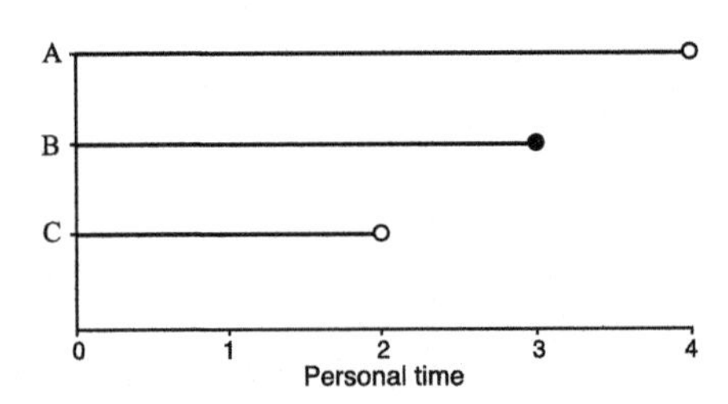

FIGURE 4.4

Open cohort study without staggered entry (case: *solid circle*; censoring: *hollow circle*).

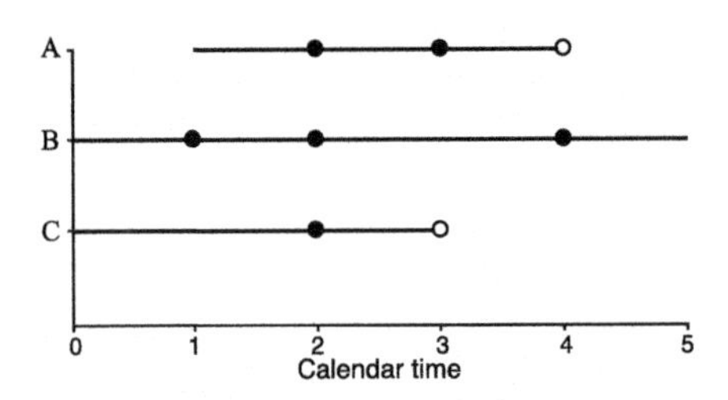

FIGURE 4.5

Open cohort study with repeated episodes permitted (episode: *solid circle*; lost to follow-up: *hollow circle*).

of a second episode only after there has been a first episode. Fig. 4.5 depicts the follow-up of three individuals, where a solid circle represents the occurrence of an episode of a given disease and a hollow circle indicates being lost to follow up. Note that the horizontal axis is calibrated in calendar time.

Consider an open cohort study based on Fig. 4.5 where becoming a case is defined to be the occurrence of a second episode. Subject A enters the cohort at time 2.0, and one year later becomes a case. Subject B enters the cohort at time 1.0 and one year later becomes a case. Subject C enters the cohort at time 2.0 and is censored at time 3.0 due to being lost to follow-up.

Now consider an open cohort study based on Fig. 4.5 where becoming a case is defined to be the occurrence of a third episode. Subject A enters the cohort at time 3.0

and is censored at time 4.0 due to being lost to follow-up. Subject B enters the cohort at time 2.0 and becomes a case at time 4.0 due to onset of a third episode. Subject C never enters the cohort.

Lastly, consider an open cohort study based on Fig. 4.5 under the assumption that episodes are independent, in the sense once a subject experiences an episode, that person is at the same risk of another episode as someone who never had an episode. We can therefore view the trajectory of subject A as the concatenation of the trajectories of three subjects who enter the cohort at time 1.0, just after time 2.0, and just after time 3.0. The first of these hypothetical subjects becomes a case at time 2.0, the second at time 3.0, and the third is censored at time 4.0 due to being lost to follow-up. In this way, all episodes of disease in Fig. 4.5 are counted.

Exponential Model

To analyze data from an open cohort study, we turn to the *exponential distribution*, a probability model that can accommodate losses to follow-up. The exponential distribution is completely characterized by a single parameter, denoted by I and referred to in epidemiology as the *incidence rate* (or *incidence density*). The *survival curve* corresponding to the exponential distribution is

$$S(t) = \exp(-It), \tag{4.7}$$

where $\exp(\cdot)$ is the exponential function corresponding to the natural logarithm. By definition, $S(t)$ is the probability of "surviving" until time t, that is, the probability of not experiencing the outcome of interest by time t. Fig. 4.6 depicts a typical exponential survival curve. It follows from (4.7) that the risk of becoming a case during a follow-up of Δ time units is

$$1 - S(\Delta) = 1 - \exp(-I\Delta). \tag{4.8}$$

For the exponential distribution to be applicable to an open cohort study, the following two conditions must be satisfied: (1) all subjects have the same incidence rate,

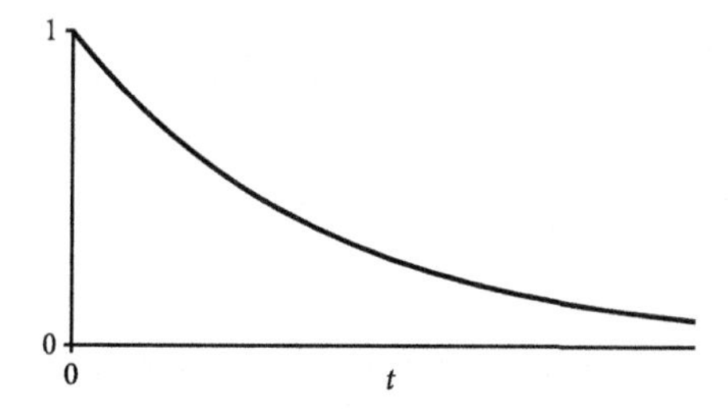

FIGURE 4.6

Exponential survival curve $\exp(-It)$.

and (2) subjects behave independently with respect to becoming a case. These conditions correspond to those required for the binomial model, and the same comments apply here.

Let us denote by a the number of incident cases. Each subject in an open cohort study (and in a closed cohort study for that matter) experiences a certain amount of personal time between the start of follow-up and either becoming a case or being censored. Let us denote by y the total amount of personal time experienced by all subjects combined. We measure y in units of *person-time*. For example, there are $y = 9.0$ person-years of follow-up in each of the cohort studies depicted in Figs. 4.1–4.4. From the perspective of the exponential distribution, both a and y are random variables. An estimator of the model parameter I is given by the random variable $\widehat{I} = a/y$. Following the convention adopted for the binomial model, we write

$$I = \frac{a}{y}. \tag{4.9}$$

The numerator of (4.9) has the units "persons", and the denominator the units "person-years", or more generally "person-time". An incidence rate therefore has the rather unusual units of "persons per person-time", or simply "per time". The magnitude of an incidence rate can be made arbitrarily large or small by choosing time units judiciously. For example, 365 per year is the same as 1 per day or 36,500 per century. The cholera mortality rates in Table 1.1 have the units "per 4 months". If the rates are divided by 4, they have the units "per month", and if the rates are multiplied by 3, they have the units "per year". Recall that a binomial risk is dimensionless with a magnitude between 0 and 1 inclusive. The preceding remarks make it clear that risks and incidence rates are distinct ways of measuring outcome in cohort studies.

Ideally the term "rate" should be reserved for indices that can be interpreted as "number of incident outcomes divided by amount of person-time". Unfortunately, this convention is not strictly adhered to in some of the standard terminology used in epidemiology and demography. For example, the case fatality rate is actually a risk, and the congenital anomaly rate is a point prevalence.

Person-time is not the only denominator used in incidence rates. Studies have shown that the threat of developing lung cancer posed by cigarette smoking depends, among other things, on the duration of smoking and the cumulative number of cigarettes smoked. We can form a composite measure called *pack-years* that combines these variables. For example, someone who has smoked one pack of cigarettes per day for ten years has accumulated ten pack-years of smoking, as has someone who has smoked two packs of cigarettes per day for five years. The corresponding incidence rate is defined to be the number of incident cases of lung cancer divided by the number of pack-years.

An important feature of the exponential model is its indifference to how person-years are accumulated. For example, one individual followed for ten years produces ten person-years, as do five individuals followed for two years. This "memoryless" property of the exponential distribution allows the follow-up of a cohort to be divided into distinct contributions. As an illustration, consider a cohort study in which cancer patients are followed until death, the aim being to determine the mortality rate corresponding to each of several stages of the disease. Over the course of follow-up, patients may progress from one stage to the next more advanced stage. They contribute deaths and person-years to the mortality rate for a given stage only when their disease is at that stage. Summing numbers of deaths across all stages gives the total number of deaths in the cohort, and similarly for person-years.

As another example, consider a cohort study in which members of a workforce in an unhealthy occupational environment are followed for a certain disease, the aim being to determine the incidence rates corresponding to risk periods of <10, 10–19 and 20+ years of exposure. Similar to the previous example, workers contribute incident cases and person-years to the incidence rate for a given risk period only when their occupational history places them in that risk period. For example, a worker enters the 20+ years category only after passing through the <10 years category and then the 10–19 years category. At the end of the first year in the 20+ years category, the worker has experienced one person-year in that category, not 21 person-years. A related consideration has to do with induction periods (see Fig. 2.1). Suppose the induction period for the disease under consideration is five years. Then someone entering the workforce is not at risk until five years have passed, at which point the worker begins to contribute person-years to the <10 years category.

Despite its many attractive features, the exponential model is based on the fiction that in the absence of censoring, there is only one way for someone to exit the cohort, and that is by experiencing the outcome of interest. Expressed differently, the exponential model assumes that all members of the cohort would eventually experience the outcome of interest if only they could be kept under observation long enough. This is realistic when the outcome is death from any cause, but otherwise such an assumption is usually difficult to justify. No matter what the outcome of interest, no matter how short the administrative length of follow-up, and no matter how vigilant the oversight, there is always the possibility that a cohort member will experience some event which removes that person from further observation prior to experiencing the outcome of interest.

An unbiased estimate of the incidence rate can be obtained using standard methods of survival analysis if censoring is *noninformative*, meaning that subjects lost to follow-up at any point during follow-up and with a given exposure profile are a simple random sample of subjects under observation in the cohort at that moment with that exposure profile.[3(p.457–470)] However, it is easy to imagine circumstances under which losses to follow-up are not a simple random sample. For example, consider an open cohort study investigating mortality after the diagnosis of a certain disease. It is conceivable that individuals who find their health failing might decide to leave the study area and seek medical treatment at some undisclosed specialty clinic. This could result in fewer deaths in the cohort coming to the attention of the investigators than would otherwise be recorded. For this reason, data from the study will tend to underestimate the true mortality risk. As another example, consider an open cohort study investigating the effect of smoking on death from lung cancer. A cohort member might be lost to follow-up because of death due to coronary heart disease, for example. Smoking is a cause of both lung cancer and coronary heart disease, and so individuals lost to follow-up due to death from coronary heart disease are not a simple random sample of those at risk of death from lung cancer.

As unrealistic as the assumption of noninformative censoring may be, once the data have been analyzed under this assumption, the findings can be evaluated both qualitatively and quantitatively based on what is known about the censoring mechanism to decide on the extent to which incidence rates may have been over- or underestimated. Alternative and more complex methods of survival analysis are available that do not rely on noninformative censoring, but instead incorporate weights conceptually similar to the design weights considered in Section 2.5.[3(p.457–470), 4(p.209–222)]

Although data from an open cohort study cannot be analyzed using the binomial model, it is possible to analyze data from a closed cohort study using the exponential model, provided the number of person-years experienced by the cohort has been recorded. The question then arises as to how estimates computed using (4.1) and (4.8) compare. By definition, R_Δ is the Δ-year risk of becoming a case based on the binomial model, and $1 - S(\Delta)$ is the corresponding probability based on the exponential model. It is shown in Appendix C that if $I\Delta$ is "small", we have the approximation

$$R_\Delta \approx 1 - S(\Delta)\,. \tag{4.10}$$

As an illustration, let $\Delta = 4$ in the closed cohort study depicted in Fig. 4.1. Then

$$R_\Delta = \frac{2}{3} = 0.666\,,$$

$$I = \frac{2}{9.0} = 0.222$$

and

$$1 - S(\Delta) = 1 - \exp\left(-\frac{2}{9.0} \times 4.0\right) = 0.589\,.$$

Since $I\Delta = 0.889$ is not "small", the approximation is not good.

Measures of Effect

Consider an open cohort that is comprised of an exposed cohort and an unexposed cohort. Suppose that over the course of follow-up there are a_1 incident cases and y_1 person-years in the exposed cohort, and a_0 incident cases and y_0 person-years in the unexposed cohort, so that $a = a_1 + a_0$ and $y = y_1 + y_0$. We assume that the exposed and unexposed cohorts are governed by exponential distributions with parameters I_1 and I_0, respectively. Consistent with (4.9), we have

$$I_1 = \frac{a_1}{y_1}$$

$$I_0 = \frac{a_0}{y_0}\,.$$

The preceding notation is summarized in Table 4.4.

Table 4.4 Data and incidence rates for open cohort study.

Exposure	Cases	Person-time	Rate
$E = 1$	a_1	y_1	I_1
$E = 0$	a_0	y_0	I_0
Overall	a	y	I

There are two basic measures of effect for open cohort studies. The *incidence rate ratio* and *incidence rate difference* are defined to be

$$IR = \frac{I_1}{I_0} = \frac{a_1/y_1}{a_0/y_0}$$

$$ID = I_1 - I_0 = \frac{a_1}{y_1} - \frac{a_0}{y_0},$$

respectively. In more sophisticated statistical analyses of data from an open cohort study, particularly those relying on regression techniques, the measure of effect is often the *hazard ratio*, which we can think of as a type of incidence rate ratio.[3]

Example: British Doctors' Study

Cigarette smoking is a cause of several diseases, in particular lung cancer and coronary heart disease. In the 1950s, the scientific evidence against cigarettes was far from conclusive, and there was vigorous debate over the role, if any, played by cigarettes in the development of these and other conditions. In 1954, Doll and Hill published the first of a series of landmark papers on the follow-up of a cohort of British physicians, which eventually extended to 50 years.[5–13] All physicians in the United Kingdom were contacted in October 1951 and asked to participate. A total of 34,445 men and 6,192 women provided information on current smoking habits (cigarettes, cigars, pipes) and agreed to be followed. Table 4.5 gives the data and coronary heart disease mortality rates for the 10-year follow-up of male physicians, where the exposed cohort is comprised of cigarette smokers, and the unexposed cohort consists of nonsmokers (of any tobacco product).[7, 14(p.205–268)] Mortality is greater in cigarette smokers compared to nonsmokers, and this is reflected in the measures of effect: $IR = 1.63$ and $ID = 0.00178$ per year. ◊

4.3 Choosing a Measure of Effect

We now consider which, if any, of the risk ratio, odds ratio and risk difference is the most desirable measure of effect for analyzing data from a closed cohort study. One of the most contentious issues revolves around the utility of these measures of

Table 4.5 Data and 10-year coronary heart disease mortality rates for male physicians in British Doctors' Study.

Smoking status	Deaths	Person-years	Rate[a]
Smoker	660	143,082	46.1
Nonsmoker	112	39,560	28.3
Overall	772	182,642	42.3

[a] *per 10,000 per year.*
Data from Reference 14.

effect for quantifying a cause-effect (etiologic) relationship on the one hand, and public health impact on the other. This can be illustrated with hypothetical examples. Suppose the risk of becoming a case is small, whether exposure is present or not, say $R_1 = 0.000003$ and $R_0 = 0.000001$. Then $RR = 3.0$, and according to usual epidemiologic practice, this is large enough to warrant further investigation of the exposure as a possible cause of the disease. On the other hand, with $RD = 0.000002$, exposure is associated with such a small increase in the risk of becoming a case that unless a large segment of the population is exposed, the impact of the disease will be negligible. From a public health perspective, the exposure is unlikely to be of great concern. Now suppose $R_1 = 0.06$ and $R_0 = 0.05$, so that $RR = 1.2$ and $RD = 0.01$. The risk ratio is likely too small to be of much interest from an etiologic point of view, but the risk difference is almost certainly of public health importance.

The above arguments have been expressed in terms of the risk ratio and risk difference, but are in essence a debate over whether the measure of effect should be *multiplicative*, such as the risk ratio, odds ratio and incidence rate ratio, or *additive*, such as the risk difference and incidence rate difference. A more fundamental consideration, and one that should be of overriding importance, is whether a measure of effect is consistent with the biology of the disease process. For example, if it is known that a set of exposures exert their influence in an additive rather than a multiplicative fashion, it would be appropriate to select the risk difference or incidence rate difference as a measure of effect in preference to the risk ratio, odds ratio or incidence rate ratio.

It might be hoped that epidemiologic data could be used to determine whether a set of exposures is operating additively, multiplicatively or in some other manner. However, the behavior of risk factors at the population level, which is the arena in which epidemiologic research operates, may not accurately reflect underlying biological processes. Currently, in the majority of epidemiologic studies, some form of multiplicative model is used. The main reason for this emphasis is almost certainly the practical consideration that in most epidemiologic investigations, the outcome variable is categorical (discrete) and the majority of statistical methods, along with most of the statistical software available to analyze such data, are based on multiplicative models.

References

1. Karlberg J, Chong DSY, Lai WYY. Do men have a higher case fatality rate of severe acute respiratory syndrome that women do? *Am J Epidemiol* 2004;**159**(3):229–31.
2. Stark CR, Mantel N. Effects of maternal age and birth order on the risk of mongolism and leukemia. *J Natl Cancer Inst* 1966;**37**(5):687–98.
3. Collett D. *Modelling survival data in medical research.* 3rd ed. Boca Raton: CRC Press; 2015.
4. Hernán MA, Robins JM. *Causal inference: what if*. Boca Raton: Chapman & Hall/CRC; 2020. https://www.hsph.harvard.edu/miguel-hernan/causal-inference-book.
5. Doll R, Hill AB. The mortality of doctors in relation to their smoking habits: a preliminary report. *Br Med J* 1954 Jun 26;**1**(4877):1451–5.
6. Doll R, Hill AB. Lung cancer and other causes of death in relation to smoking: a second report on the mortality of British Doctors. *Br Med J* 1956 Nov 10;**2**(5001):1071–81.
7. Doll R, Hill AB. Mortality in relation to smoking: ten years' observations of British doctors. *Br Med J* 1964 May 30;**1**(5395):1399–410.
8. Doll R, Hill AB. Mortality in relation to smoking: ten years' observations of British doctors. *Br Med J* 1964 Jun 6;**1**(5396):1460–7.
9. Doll R, Peto R. Mortality in relation to smoking: 20 years' observations on male British doctors. *Br Med J* 1976 Dec 25;**2**(6051):1525–36.
10. Doll R, Gray R, Hafner B, et al. Mortality in relation to smoking: 22 years' observations on female British doctors. *Br Med J* 1980 Apr 5;**280**(6219):967–71.
11. Doll R, Peto R, Wheatley K, et al. Mortality in relation to smoking: 40 years' observations on male British doctors. *BMJ* 1994 Oct 8;**309**(6959):901–11.
12. Doll R, Peto R, Boreham J, et al. Mortality in relation to smoking: 50 years' observations on male British doctors. *BMJ* 2004 June 26;**328**(7455):1519–28.
13. Doll R, Peto R, Wheatley K, et al. Mortality from cancer in relation to smoking: 50 years' observations on male British doctors. *Br J Cancer* 2005;**92**(3):426–9.
14. Doll R, Hill AB. Mortality of British doctors in relation to smoking: observations on coronary thrombosis. In: Haenszel W, editor. *Epidemiological approaches to the study of cancer and other chronic diseases. National Cancer Institute monograph 19.* Bethesda: U.S. Department of Health, Education, and Welfare; 1966.

CHAPTER

Populations

5

The term *population* is used throughout epidemiology in a variety of contexts. For example, the subjects in an epidemiologic study are often referred to as the study population. The definitions of source population and target population were given in Section 2.6. To the extent that established conventions permit, we will restrict our use of the term population to mean the residents of a geographically-defined entity, such as a city or a country.

The material presented in this chapter is a hybrid of methods used by epidemiologists and demographers to study health-related phenomena in populations. An epidemiologic perspective can be introduced by viewing a population as an open cohort with staggered entry that is observed over a (calendar) time period of, say, Δ years.

5.1 Incidence Rates

Let us denote by A the number of incident cases of a disease of interest occurring in a population during an observation period, and by Y the number of person-years experienced during the observation period by those individuals in the population at risk of becoming a case. We assume that the disease can occur more than once in a lifetime and that everyone in the population who is not a case at any given time during the observation period is at risk of becoming one. This corresponds to the third interpretation of Fig. 4.5 in which episodes of disease are viewed as independent. The *crude incidence rate* of the disease for the observation period (more precisely, the *crude Δ-year incidence rate*) is defined to be [1,2]

$$I = \frac{A}{Y}, \tag{5.1}$$

which has the units "cases per person-year", or equivalently "per year". An important difference from the open cohort model described in Section 4.2 is that there are generally many diseases operating simultaneously in a population, with the disease of interest being just one of them. Accordingly, I is the incidence rate of the disease of interest "in the presence of other diseases". Ordinarily the value of A would come from a vital statistics database or a disease registry. Obtaining a value of Y presents more of a challenge. We present two ways of approaching this problem.

Epidemiologic Methods. https://doi.org/10.1016/B978-0-44-318780-3.00011-7
Copyright © 2023 Elsevier Inc. All rights reserved.

Let M' denote the number of individuals in the population at the midpoint of the observation period who are at risk of becoming a case at that moment. We take M' to be an estimate of the "average number" of individuals in the population at any time during the observation period who are at risk of becoming a case. Since the length of the observation period is Δ years, we have the approximation

$$Y \approx M'\Delta\,. \tag{5.2}$$

Let us denote by N' the number of individuals in the population at the midpoint of the observation period, and by C' the number of prevalent cases in the population at that time. Then $P' = C'/N'$ is the corresponding point prevalence. By assumption, everyone in the population who is not a case at the midpoint of the observation period is at risk of becoming one. It follows that

$$M' = N' - C' = (1 - P')N'\,. \tag{5.3}$$

Combining (5.1)–(5.3) yields

$$I \approx \frac{A}{M'\Delta} = \frac{A}{(1 - P')N'\Delta}\,. \tag{5.4}$$

The shorter the observation period, the better the approximation in (5.4) is likely to be. The value of N' would usually be obtained from a population census conducted by a government agency, and the value of P' from a prevalence study, again conducted by a government agency or perhaps by a university or other research organization.

As an alternative to the above approximations, let us assume that the number of individuals in the population during the observation period and the number of prevalent cases in the population during the observation period are independent of time. A population satisfying these assumptions (or something similar depending on the population under consideration) is said to be *stationary*. We adopt the above notation but now drop the prime ($'$). Accordingly, there are N individuals in the population at any time during the observation period, M individuals in the population at any time during the observation period at risk of becoming a case, and C prevalent cases in the population at any time during the observation period. It follows that the point prevalence $P = C/N$ is also independent of time and that $Y = M\Delta$. We then have from (5.1) and (5.3) that

$$M = N - C = (1 - P)N \tag{5.5}$$

and

$$I = \frac{A}{M\Delta} = \frac{A}{(1 - P)N\Delta}\,, \tag{5.6}$$

where we note that the first relation in (5.6) is an equality, not an approximation. When the "disease of interest" is death from all causes or a specific cause, the crude incidence rate is called the *crude mortality rate* (or *crude death rate*). In this situation, the number of prevalent cases is zero and the entire population is at risk.

As a practical matter, the choice between the above two approaches to estimating I is more a matter of style than substance. Regardless of whether (5.4) or (5.6) is used, computations ultimately rely on the same census, incidence and prevalence data. For convenience of notation and, more importantly, because it fits well with discussions in later chapters, we now assume that the population under observation is stationary.

Demographic characteristics, such as age, gender and ethnicity, taken singly or in combination, can be used to stratify the population. For simplicity, we focus on age as the sole stratifying variable. Suppose the age range in the population has been divided into $k+1$ nonoverlapping age groups and that the population in each age group is stationary. We add a subscript i to the above notation to indicate age group i, for $i = 0, 1, \ldots, k$. (The indexing begins with $k = 0$ to conform to a convention established in Chapter 6.) Accordingly, A_i is the number of incident cases occurring in age group i during the observation period, and M_i is the number of individuals at risk in age group i at any time during the observation period.

The individuals at risk in age group i cumulatively experience $M_i \Delta$ person-years. The *age-specific incidence rate* for age group i is defined to be

$$I_i = \frac{A_i}{M_i \Delta}. \tag{5.7}$$

It is shown in Appendix D that the crude incidence rate is a weighted average of age-specific incidence rates:

$$I = \sum_{i=0}^{k} \left(\frac{M_i}{M} \right) I_i. \tag{5.8}$$

We note that the proportions

$$\frac{M_0 \Delta}{M \Delta}, \ldots, \frac{M_i \Delta}{M \Delta}, \ldots, \frac{M_k \Delta}{M \Delta},$$

which can be expressed as

$$\frac{M_0}{M}, \ldots, \frac{M_i}{M}, \ldots, \frac{M_k}{M},$$

give the person-years distribution of the at-risk population for the observation period, or equivalently the age distribution of the at-risk population at any point during the observation period.

5.2 Direct Standardization

Let us now consider two stationary populations, referred to as population 1 and population 0. In keeping with Chapter 4, we can think of population 1 as being "exposed" and population 0 as being "unexposed". We are interested in comparing the crude incidence rates in the two populations. The *(crude) incidence rate ratio*, defined to be

$$IR = \frac{I_1}{I_0}, \tag{5.9}$$

can be used for this purpose. When the "disease of interest" is death from all causes or a specific cause, the crude incidence rate ratio is called the *(crude) mortality ratio* and denoted here by MR.

Example: Snow Cholera Study

Returning to the Snow cholera study discussed in Section 1.4, the numerators of the rates in Table 1.1 are the numbers of cholera deaths in the 4-month period July to October 1854, and the denominators are population estimates for some time in 1851. Let us assume that the population estimates for 1851 are applicable to the 4-month period in 1854. Taking the units of time to be months and setting $\Delta = 4$, we have from (5.6) that the crude 4-month cholera mortality rates for clients of the Southwark and Vauxhall Company (population 1) and the Lambeth Company (population 0) are

$$I_1 = \frac{4,093}{266,516 \times 4} = \frac{153.6 \times 10^{-4}}{4} \text{ per month}$$

and

$$I_0 = \frac{461}{173,748 \times 4} = \frac{26.5 \times 10^{-4}}{4} \text{ per month},$$

respectively. This yields a crude mortality ratio of

$$MR = \frac{(153.6 \times 10^{-4})/4}{(26.5 \times 10^{-4})/4} = 5.80. \qquad \diamond$$

Example: Alaska-Florida Mortality Comparison

In this example, we compare all-causes mortality for males in Alaska (population 1) and Florida (population 0). The data were downloaded from the Compressed Mortality File 1999–2016 at the Centers for Disease Control and Prevention WONDER website.[3] Table 5.1(a) gives numbers of deaths from all causes and mid-year population counts for Alaska and Florida males in 2015, by age group. Table 5.1(b) gives the crude and age-specific mortality rates and the corresponding crude and age-specific mortality ratios. Table 5.1(c) gives the age distribution of males in Alaska, Florida and United States in 2015.

From Table 5.1(b), the crude mortality rates for Alaska and Florida are $I_1 = 78.0 \times 10^{-4}$ per year and $I_0 = 122.0 \times 10^{-4}$ per year. The corresponding crude mortality ratio is

$$MR = \frac{78.0 \times 10^{-4}}{122.0 \times 10^{-4}} = 0.64.$$

Table 5.1(a) Deaths from all causes and population counts in Alaska and Florida males, 2015.

Age group	Alaska		Florida	
	Deaths	Population	Deaths	Population
15–24	86	59,445	1,425	1,268,534
25–34	126	62,666	2,244	1,318,152
35–44	130	47,831	2,851	1,210,850
45–54	273	50,550	6,927	1,346,665
55–64	495	49,168	14,718	1,249,650
65–74	493	25,665	20,630	1,012,855
75–84	475	8,218	24,860	551,087
85+	307	2,290	25,936	207,656
Overall	2,385	305,833	99,591	8,165,449

Data from Reference 3.

Table 5.1(b) Mortality rates and mortality ratios for all causes in Alaska and Florida males, 2015.

Age group	Mortality rate[a]		Mortality ratio
	Alaska	Florida	
15–24	14.5	11.2	1.29
25–34	20.1	17.0	1.18
35–44	27.2	23.5	1.15
45–54	54.0	51.4	1.05
55–64	100.7	117.8	0.85
65–74	192.1	203.7	0.94
75–84	578.0	451.1	1.28
85+	1,340.6	1,249.0	1.07
Overall	78.0	122.0	0.64

[a] *per 10,000 per year.*
Data from Reference 3.

Based on crude mortality rates, mortality in Alaska appears to be less than that in Florida. However, examining the age-specific mortality rates and age-specific mortality ratios in Table 5.1(b), we observe that for every age group except 55–64 and 65–74, the age-specific mortality rate in Alaska is greater than that in Florida. From the perspective of age-specific mortality rates, mortality in Alaska appears to be greater than in Florida.

Table 5.1(c) Age distribution (percent) of Alaska, Florida and United States males, 2015.

Age group	Alaska	Florida	U.S.
15–24	19.4	15.5	17.7
25–34	20.5	16.1	17.6
35–44	15.6	14.8	15.9
45–54	16.5	16.5	16.8
55–64	16.1	15.3	15.5
65–74	8.4	12.4	10.1
75–84	2.7	6.8	4.7
85+	0.8	2.5	1.7

Data from Reference 3.

What accounts for this apparent paradox? The answer lies in the differing age distributions of the two populations. As can be seen from Table 5.1(c), the age distribution of Florida is skewed toward the elderly compared to the age distribution of Alaska. For example, approximately one-fifth of the Florida population is 65+ years old, whereas for Alaska it is approximately one-tenth. Whether you live in Alaska or Florida, the oldest age groups have by far the largest mortality rates. As a result, relatively more deaths are contributed to the numerator of the crude mortality rate for Florida than for Alaska. ◊

We now return to the general setting of population 1 and population 0. In an obvious notation, the age-specific incidence rates for population 1 and population 0 are defined to be

$$I_{1i} = \frac{A_{1i}}{M_{1i}\Delta}$$
$$I_{0i} = \frac{A_{0i}}{M_{0i}\Delta}, \tag{5.10}$$

respectively. It follows from (5.8) that I_1 and I_0 can be expressed as

$$I_1 = \sum_{i=0}^{k} \left(\frac{M_{1i}}{M_1}\right) I_{1i}$$
$$I_0 = \sum_{i=0}^{k} \left(\frac{M_{0i}}{M_0}\right) I_{0i}. \tag{5.11}$$

This makes it clear how person-years distributions (equivalently, age distributions) that differ across populations can distort the comparison of crude incidence rates. One approach to the problem introduced by differing person-years distributions would be to avoid crude incidence rates altogether and consider only age-specific incidence rates. This has the potential to uncover interesting patterns across age groups, but it would nevertheless be convenient to have an overall index that summarizes incidence across age groups while accounting for differences in person-years distributions.

To that end, we modify the crude incidence rates for population 1 and population 0 by replacing the person-years distributions of those at risk in these populations with the person-years distribution of those at risk in a judiciously chosen population, which we refer to as the *standard population*. The choice of standard population for any given application is a matter of preference, but usually it is not difficult to identity one or more reasonable possibilities. Let population $*$ be such a standard population. The *standardized incidence rate* (more precisely, the *age-standardized incidence rate*) for population 1, with population $*$ as the standard population, is defined to be

$$I_{1(*)} = \sum_{i=0}^{k} \left(\frac{M_{*i}}{M_{*}} \right) I_{1i} \,. \tag{5.12}$$

When the "disease of interest" is death from all causes or a specific cause, $I_{1(*)}$ is called the *standardized mortality rate* (or *standardized death rate*). It is clear from (5.11) and (5.12) that

$$I_{1(1)} = I_1 \,. \tag{5.13}$$

That is, the standardized incidence rate for a population, where the population itself is the standard population, is just the crude incidence rate for that population.

It is tempting to interpret $I_{1(*)}$ as the crude incidence rate that population 1 would have experienced during the observation period if those at risk in population 1 had had the person-years distribution of those at risk in population $*$, but retained their age-specific incidence rates. A problem with this interpretation is that the person-years distribution of those at risk in population 1 is determined in part by those same age-specific incidence rates.[4] Even if it were possible to arrange for those at risk in population 1 to suddenly have the age distribution of those at risk in population $*$, immediately after such a demographic makeover, there would be a drift away from the newly-established age distribution and with that a shift away from the hoped-for person-years distribution. So the preceding interpretation of $I_{1(*)}$ is flawed. This detracts from the intuitive appeal of $I_{1(*)}$, but not its utility—and popularity—as a descriptive measure of incidence.

We are now in a position to present an alternative to the crude incidence rate ratio, and one that accounts for differences in the person-years distributions of population 1 and population 0. The *directly standardized incidence rate ratio* is denoted here by dIR and defined to be

$$dIR = \frac{I_{1(*)}}{I_{0(*)}} \,. \tag{5.14}$$

The data needed to compute dIR are the age-specific numbers of incident cases for populations 1 and 0, and the age-specific population counts of those at risk in populations 1, 0 and $*$. When the "disease of interest" is death from all causes or a specific cause, the directly standardized incidence rate ratio is called the *directly standardized mortality ratio* and denoted here by dMR.

Example: Alaska-Florida Mortality Comparison

Returning to the Alaska-Florida comparison, let us choose the male population of the United States in 2015 to be the standard population (population $*$). From Tables 5.1(b) and 5.1(c), we have

$$\begin{aligned} I_{1(*)} &= (17.7 \times 10^{-2} \times 14.5 \times 10^{-4}) + \cdots + (1.7 \times 10^{-2} \times 1,340.6 \times 10^{-4}) \\ &= 104.9 \times 10^{-4} \text{ per year} \end{aligned}$$

and

$$\begin{aligned} I_{0(*)} &= (17.7 \times 10^{-2} \times 11.2 \times 10^{-4}) + \cdots + (1.7 \times 10^{-2} \times 1,249.0 \times 10^{-4}) \\ &= 99.0 \times 10^{-4} \text{ per year}\,. \end{aligned}$$

The directly standardized mortality ratio is

$$dMR = \frac{104.9 \times 10^{-4}}{99.0 \times 10^{-4}} = 1.06\,.$$

In contrast to the analysis based on crude mortality rates, which yielded a crude mortality ratio of $MR = 0.64$, we now find that mortality in Alaska is slightly increased compared to Florida after accounting for differences in age distributions. ◊

Example: Lung Cancer Mortality Trends

The Alaska-Florida example shows how direct standardization can be employed in a comparison of two populations for a given year. Direct standardization can also be used to compare a population with itself over time. We illustrate the approach using age-specific lung cancer mortality rates for white males in the United States by 5-year age groups (50–54, 55–64, ..., 75–79) and 5-year time periods (1946–1950, 1951–1955, ..., 1971–1975). The data are taken from a study to be considered in greater detail below.[5] Taking $\Delta = 5$, we view the white male population of the United States during 1946–1975 as a sequence of six stationary populations, one for each 5-year time period. Fig. 5.1 shows a graph of the corresponding age-standardized lung cancer mortality rates over time, with the white male population of the United States in the 50–79 age group in 1975 as the standard population. For purposes of constructing the graph, mortality rates have been assigned to the midpoint of their time period. ◊

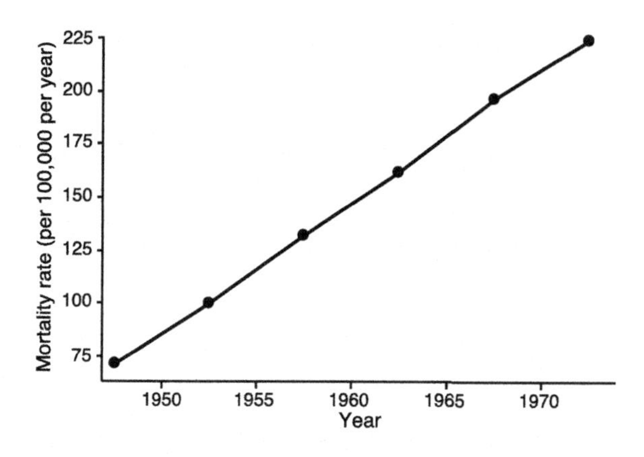

FIGURE 5.1

Age-standardized lung cancer mortality rates for U.S. white males, 50–79 years old, 1946–1975; standard population is U.S. white males, 1975.
Data from Reference 5.

5.3 Indirect Standardization

We now present an alternative to direct standardization for comparing incidence in two stationary populations. Continuing with population 1 and population 0, the *indirectly standardized incidence rate ratio* is denoted by sIR and defined to be the special case of the directly standardized incidence rate ratio in which population 1 is taken to be the standard population. We have from (5.13) and (5.14) that

$$sIR = \frac{I_{1(1)}}{I_{0(1)}} = \frac{I_1}{I_{0(1)}}, \tag{5.15}$$

where, based on (5.12),

$$I_{0(1)} = \sum_{i=0}^{k} \left(\frac{M_{1i}}{M_1} \right) I_{0i} \tag{5.16}$$

is the standardized incidence rate for population 0, with population 1 as the standard population. It is shown in Appendix D that sIR is a weighted average of age-specific incidence rate ratios. When the "disease of interest" is death from all causes or a specific cause, the indirectly standardized incidence rate ratio is called the *standardized mortality ratio* and denoted by sMR (or SMR).

Age-specific population counts are routinely published by government agencies, but numbers of incident cases are often less readily available. The data needed to compute the indirectly standardized incidence rate ratio are the total number of incident cases for population 1, the age-specific numbers of incident cases for population 0, and the age-specific population counts of those at risk in populations 1 and 0. This means that, unlike the directly standardized incidence rate ratio, the indirectly standardized incidence rate ratio can be computed even when age-specific numbers of incident cases for population 1 are unavailable.

We now approach the indirectly standardized incidence rate ratio from a somewhat different perspective. By definition, A_1 is the *observed number* of incident cases in population 1 during the observation period. The *expected number* of incident cases in population 1 during the observation period is defined to be

$$A_{0(1)} = \sum_{i=0}^{k} I_{0i} M_{1i} \Delta \, . \tag{5.17}$$

Despite being called an "expected" number, $A_{0(1)}$ is not an expected value in the statistical sense of the term. Similar to before, we are inclined to interpret $A_{0(1)}$ as the number of incident cases that would have occurred in population 1 during the observation period if it had had the age-specific incidence rates of population 0, but retained the person-years distribution of those at risk. However, we run into a problem of interpretation similar to that encountered with the standardized incidence rate.

It is shown in Appendix D that[6(p.269–272)]

$$sIR = \frac{A_1}{A_{0(1)}} \, . \tag{5.18}$$

That is, the indirectly standardized incidence rate ratio equals the observed number of incident cases in population 1 divided by the expected number. This is the classical way of defining this index, especially in studies of mortality.

A confusing aspect of the epidemiologic and demographic literature in this area is that it is population 0, not population 1, that is usually said to be the standard population in (5.17) and (5.18). This conflicts with our earlier usage of the expression standard population because in (5.17) and (5.18), population 0 contributes its age-specific incidence rates to the computation, not its person-years distribution. This can lead to errors when interpreting indirectly standardized incidence rate ratios.[7] To illustrate the potential problem, suppose we wish to compare incidence in populations 1, 2 and 3 to population 0 using indirect standardization. Denoting the observed numbers of incident cases by A_1, A_2 and A_3, respectively, the indirectly standardized incidence rate ratios are

$$\frac{A_1}{A_{0(1)}}, \frac{A_2}{A_{0(2)}}, \frac{A_3}{A_{0(3)}} \, .$$

We may be inclined to compare the above quotients to each other. But this would be inappropriate because each of them has been standardized to a different person-years distribution. To avoid confusion, we will use the expression standard population exclusively to refer to a population that provides its person-years distribution to standardization computations.

Example: Alaska-Florida Mortality Comparison

Continuing with the Alaska-Florida example, we have $I_1 = 78.0 \times 10^{-4}$ per year and

$$I_{0(1)} = (19.4 \times 10^{-2} \times 11.2 \times 10^{-4}) + \cdots + (0.8 \times 10^{-2} \times 1,249.0 \times 10^{-4})$$
$$= 75.4 \times 10^{-4} \text{ per year},$$

hence

$$sMR = \frac{78.0 \times 10^{-4}}{75.4 \times 10^{-4}} = 1.04 .$$

This shows that mortality in Alaska is slightly increased compared to Florida after accounting for differences in age distributions. Recall that the directly standardized mortality ratio computed earlier was $dMR = 1.06$, which led to the same conclusion. ◊

5.4 Age-Period-Cohort Analysis

Consider a geographically-defined population that is under observation for the occurrence of a disease over an extended time period, say several decades. Most diseases exhibit a pattern of incidence and mortality that varies with age, which is referred to as an *age effect*. A factor that impacts a population as a whole over a relatively short time frame is called a *period effect*. For example, a period effect could result from a natural disaster, such as a flood or influenza epidemic. A factor that exerts its influence on certain members of the population who then carry the consequences forward in time is said to be a *cohort effect*. As an example, a cohort effect might be seen following a nutrition program for young children, who then have improved resistance to disease and other benefits later in life. The preceding examples are fairly straightforward, but even here it may be difficult to separate period from cohort effects. For instance, the attention surrounding the introduction of a nutrition program for young children could generate a meaningful, but short-lived, improvement in the eating habits of the general population. So there might be a period effect as well as a cohort effect. A data analysis that attempts to tease apart the three types of effects in a population is said to be an *age-period-cohort analysis*.[5,8]

Let us turn our attention to a detailed example based on data considered in connection with Fig. 5.1.[5] Table 5.2 gives the lung cancer mortality rates for U.S. white males by 5-year age groups (50–54, 55–59, ..., 75–79) and 5-year time periods (1946–1950, 1951–1955, ..., 1971–1975). Before proceeding, a few remarks are in order regarding the notation for age groups and time periods. The 50–54 age group, for example, includes all individuals who have had a 50th birthday but not a 55th birthday. In continuous time, this includes all ages greater than or equal to 50.0 and less than 55.0, an interval that is five years long. Someone alive in 1946 is in the 1946th year of the Gregorian calendar, which in continuous time begins at 1945.0

and ends just before 1946.0. Similarly, in continuous time, the 1946–1950 time period begins at 1945.0 and ends just before 1950.0, an interval which is five years long.

Let us examine the rows and columns of Table 5.2 for patterns. For each of the columns, the rates increase with time period. We interpret the phenomenon that takes one column of Table 5.2 to the next as an *age effect*. Fig. 5.2 is the associated graph, where the curves correspond to the columns of Table 5.2. For clarity, the age groups and time periods are labeled as in Table 5.2 but are to be imagined in continuous variable terms. Thus, the 50–54 age group can be thought of as the point 52.5, the midpoint of 50–54. Likewise, the 1946–1950 time period can be thought of as the point 1947.5. The stacked appearance of (most of) the curves is evidence of age effects.

For each of the rows of Table 5.2, the rates for the most part increase with age group. We interpret the phenomenon that takes one row to the next as a *period effect*. Fig. 5.3 is the associated graph, where the curves correspond to the rows of Table 5.2. The stacked appearance of the curves is evidence of period effects. Fig. 5.1 is consistent with this finding.

We can create birth cohorts by following individuals "along the diagonal" of Table 5.2, which gives Table 5.3. Each birth cohort is labeled according to the years during which its members were born. To illustrate, someone who was in the 50–54 age group at some point in 1946–1950 and lived another five years would then have been in the 55–59 age group at the corresponding point in 1951–1955. The members of this birth cohort were born during the time period 1891–1900, which is ten years long. Similarly, someone who was in the 50–54 age group at some point in 1951–1955 and lived another five years would then have been in the 55–59 age group at the corresponding point in 1956–1960. The members of this cohort were born during the time period 1896–1905, which again is ten years long. Note that the time periods during which the preceding two birth cohorts were born, that is, 1891–1900 and 1896–1905, have the overlap 1896–1900, which is five years long. We assume

Table 5.2 Lung cancer mortality rates (per 100,000 per year) by time period and age group, U.S. white males.

Time period	Age group					
	50–54	55–59	60–64	65–69	70–74	75–79
1946–1950	43.3	68.2	87.4	87.5	85.6	77.7
1951–1955	53.8	88.4	120.9	136.2	128.9	115.2
1956–1960	64.3	108.9	158.3	189.1	185.0	160.8
1961–1965	72.1	121.9	189.4	240.3	245.2	224.7
1966–1970	81.4	139.5	218.8	289.1	322.4	318.7
1971–1975	86.4	151.9	238.5	325.1	390.9	415.4

Data from Reference 5.

that this overlap can be ignored, allowing us to treat the birth cohorts as independent. We also assume that the net effect of in-migration and out-migration in the U.S. white male population during 1946–1975 was such that the mortality rates in each age group and time period were not affected. The only justification offered here for these assumptions is that they are necessary for what follows. With this preparation, we can treat the birth cohorts in Table 5.3 as cohorts in the sense of Chapter 4.

For each of the rows of Table 5.3, the rates increase with age group. We interpret the phenomenon that takes one row of Table 5.3 to the next as a *cohort effect*. Fig. 5.4

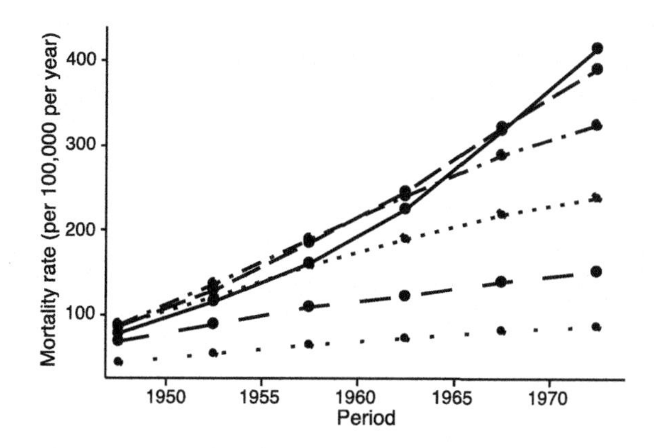

FIGURE 5.2

Columns of Table 5.2: age effects (50–54: *spacedot*; 55-59: *spacedash*; 60–64: *dot*; 65–69: *dashdot*; 70–74: *dash*; 75-79: *solid*).
Data from Reference 5.

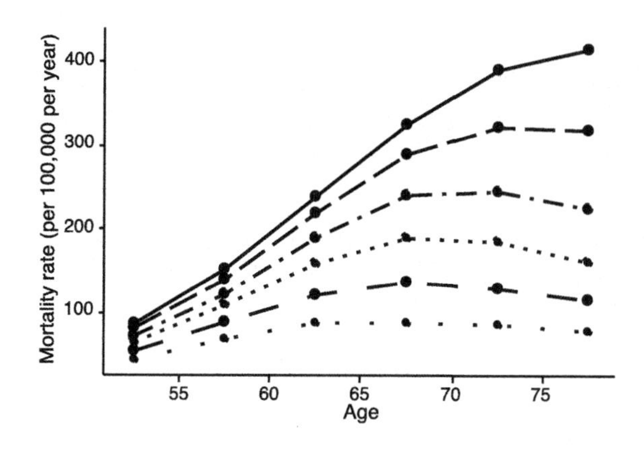

FIGURE 5.3

Rows of Table 5.2: period effects (1946–1950: *spacedot*; 1951–1955: *spacedash*; 1956-1960: *dot*; 1961–1965: *dashdot*; 1966–1970: *dash*; 1971–1975: *solid*).
Data from Reference 5.

Table 5.3 Lung cancer mortality rates (per 100,000 per year) by birth cohort and age group, U.S. white males.

Birth cohort	Age group					
	50–54	55–59	60–64	65–69	70–74	75–79
1866–1875	–	–	–	–	–	77.7
1871–1880	–	–	–	–	85.6	115.2
1876–1885	–	–	–	87.5	128.9	160.8
1881–1890	–	–	87.4	136.2	185.0	224.7
1886–1895	–	68.2	120.9	189.1	245.2	318.7
1891–1900	43.3	88.4	158.3	240.3	322.4	415.4
1896–1905	53.8	108.9	189.4	289.1	390.9	–
1901–1910	64.3	121.9	218.8	325.1	–	–
1906–1915	72.1	139.5	238.5	–	–	–
1911–1920	81.4	151.9	–	–	–	–
1916–1925	86.4	–	–	–	–	–

Data from Reference 5.

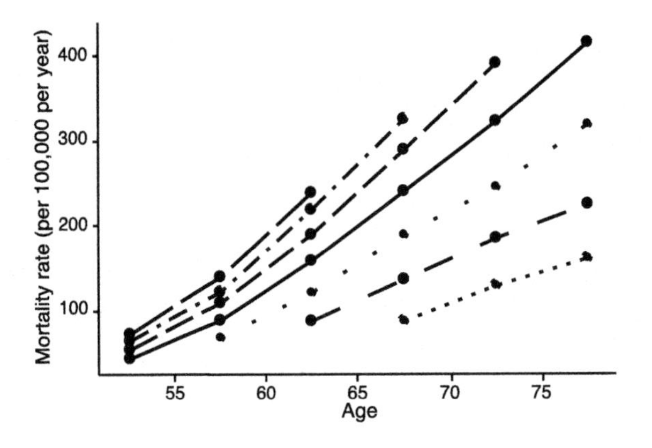

FIGURE 5.4

Rows of Table 5.3: cohort effects (1876–1885: *dot*; 1881–1990: *spacedash*; 1886–1895: *spacedot*; 1891–1900: *solid*; 1896–1905: *dash*; 1901–1910: *dashdot*; 1906–1915: *long-dash*).
Data from Reference 5.

is the associated graph, where the curves correspond to selected rows of Table 5.3. The stacked appearance of the curves is evidence of cohort effects.

The findings in Figs. 5.3 and 5.4 raise interesting questions about the relative contributions of period effects and cohort effects to the pattern of mortality from lung cancer. Let us imagine for a moment much simpler versions of Figs. 5.3 and

5.4. Suppose all the curves in Fig. 5.3 were overlaid except the curve for 1946–1950, which was elevated above the rest. The corresponding Fig. 5.4 would also be a series of overlaid curves, except that each one would have a "tooth" pointing upwards, so that taken together they would present a saw-toothed appearance. With graphs such as this, it would be easy to conclude that there had been a period effect during 1946–1950. On the other hand, suppose all the curves in Fig. 5.4 were overlaid except the curve for 1891–1900, which was elevated above the rest. We would then find a saw-toothed appearance in the corresponding Fig. 5.3 and conclude from these graphs that there had been a cohort effect in 1891–1900.

In practice, the patterns exhibited by Figs. 5.2–5.4 are typical of what is commonly seen with chronic diseases: an increasing or decreasing trend across age groups, an increasing or decreasing trend across time periods, and an increasing or decreasing trend across birth cohorts. How do we tease apart the relative contributions of the period and cohort effects? It might be hoped that the problem could be resolved with the aid of multivariate models that have terms for age effects, period effects and cohort effects. Let us pursue this possibility in the context of Figs. 5.2–5.4.

In what follows, we set aside the stationary assumption and treat age at birth ("age"), time ("period") and birth cohort ("cohort") as continuous variables, denoted by x, y and z, respectively. It is clear that someone who is x years old at time y was born at time $y - x$. We therefore have the fundamental identity

$$z = y - x. \tag{5.19}$$

To construct the desired multivariate model, let $I_{x,y}$ be the lung cancer mortality rate for someone who is age x at time y. Restricting Table 5.2 to the four age groups 50–54, 55–59, 60–64, 65–69 and four time periods 1956–1960, 1961–1965, 1966–1970, 1971–1975, the pattern of the corresponding curves in Fig. 5.2 suggests the exponential model

$$I_{x,y} = \exp(b + ax + py), \tag{5.20}$$

where b, a and p are constants to be estimated from the data. Setting $x = y = 0$, we find that the baseline mortality rate is $I_{0,0} = \exp(b)$. For a given value of y, each unit increase in x increases mortality rates by a factor $I_{x+1,y}/I_{x,y} = \exp(a)$. Thinking in terms of the columns of Table 5.2, $\exp(a)$ can be interpreted as an age effect. Similarly, for a given value of x, each unit increase in y increases mortality rates by a factor $I_{x,y+1}/I_{x,y} = \exp(p)$. Now thinking in terms of the rows of Table 5.2, $\exp(p)$ can be interpreted as a period effect.

Let $J_{x,z}$ be the lung cancer mortality rate for someone who is age x and was born at time z. It follows from (5.19) that the mortality rate for someone who is age x and was born at time z is precisely the mortality rate for someone who is age x at time $y = x + z$. We therefore have the identity

$$J_{x,z} = I_{x,x+z}\,. \tag{5.21}$$

Then (5.20) and (5.21) give

$$J_{x,z} = \exp[b + ax + p(x + z)]$$
$$= \exp[b + (a + p)x + pz].$$

We find that $J_{0,0} = \exp(b)$ is the baseline mortality rate, as before. For a given value of x, each unit increase in z increases mortality rates by a factor $J_{x,z+1}/J_{x,z} = \exp(p)$. Thinking in terms of the rows of Table 5.3, we find that $\exp(p)$ can be interpreted as a cohort effect.

The above observations show that in age-period-cohort analyses based on models (5.20) and (5.21), a period effect can manifest as a cohort effect, and conversely. This ambiguity cannot be resolved by considering more complicated mathematical models. The difficulty lies in the fact that, due to the identity $z = y - x$, age effects, period effects and cohort effects are inextricably linked. This is referred to as the *identification problem* for age-period-cohort models. In order to separate the three effects, it is necessary to have an additional equation relating the age, period and cohort variables. The choice of equation rests on prior knowledge of the disease and the population; it is not a statistical issue. In most applications it is unusual for such an equation to be available, and so, unfortunately, age-period-cohort models cannot be relied upon by themselves to dissect the three time-related effects.

The usual approach to analyzing age-period-cohort data is to make use of graphs, such as those above, multivariate models and substantive knowledge to arrive at a determination as to the relative importance of age effects, period effects and cohort effects on incidence and mortality in the population.

5.5 Prevalence = Incidence x Duration

Let us return to the setting of a single stationary population. We define the *prevalence pool* to be the "state" of being a prevalent case. That is, the prevalence pool at a given time consists of those members of the population who are prevalent cases at that time. There are $C = PN$ people in the prevalence pool at any point during the observation period, and so the prevalence pool cumulatively experiences $C\Delta = PN\Delta$ person-years during the observation period. We now assume the prevalence pool does not experience in-migration or out-migration. Since there is no in-migration, the only way to enter the prevalence pool is from that part of the population outside the prevalence pool. We have from (5.5) and (5.6) that the number of entrances into the prevalence pool during the observation period, that is, the number of incident cases, is $IM\Delta = I(1 - P)N\Delta$. Since there are $C = PN$ individuals in the prevalence pool at the start of the observation period, a total of $PN + I(1 - P)N\Delta$ partial or complete episodes of the disease take place in the prevalence pool during the observation period.

Let us denote by $\overline{D}_{\Delta}$ the average number of years spent in the prevalence pool during the observation period for each partial or complete episode of the disease.

With $PN\Delta$ person-years and $PN + I(1-P)N\Delta$ partial or complete episodes, we find that

$$\begin{aligned}\overline{D}_{\Delta} &= \frac{PN\Delta}{PN + I(1-P)N\Delta} \\ &= \frac{P}{P/\Delta + I(1-P)}.\end{aligned} \tag{5.22}$$

Since there is no out-migration, the only way to exit the prevalence pool is by recovery from the disease, or death from some cause. It follows that as Δ gets larger, $\overline{D}_{\Delta}$ more closely approximates the average duration of a complete episode of the disease, which we denote by $\overline{D}$. In the limit, the first term in the denominator of the second line of (5.22) goes to zero and we have[1,9,10]

$$\frac{P}{1-P} = I\overline{D}. \tag{5.23}$$

Using (3.2), we can rewrite (5.23) in terms of the point prevalence odds as

$$PO = I\overline{D}. \tag{5.24}$$

When P is "small", (3.3) and (5.24) yield the approximation

$$P \approx I\overline{D},$$

which is a version the classic identity Prevalence = Incidence × Duration.

5.6 Cross-Sectional Studies

For the moment, let us consider a population that is not necessarily stationary. Data collected for a prevalence study of the population at time t can be used to investigate the association between an exposure and a disease, giving what is called a *cross-sectional study*. Suppose the members of the population at any given time can be categorized as either exposed or unexposed. The *point prevalence ratio* and *point prevalence odds ratio* at time t are defined to be[1,10]

$$PR = \frac{P_1}{P_0} \tag{5.25}$$

$$POR = \frac{PO_1}{PO_0}, \tag{5.26}$$

respectively, where P_1 and P_0 are prevalences, PO_1 and PO_2 are prevalence odds, and the subscripts have the usual meaning. The *period prevalence ratio* and *period prevalence odds ratio* are defined similarly. If P_1 and P_0 are "small", (3.3), (5.25) and (5.26) combine to give the approximation[1]

$$POR \approx PR.$$

Let us assume that the exposed and unexposed populations are each stationary and that their prevalence pools do not experience in-migration or out-migration. It follows from (5.9), (5.24) and (5.26) that[1,10]

$$POR = \frac{I_1 \overline{D}_1}{I_0 \overline{D}_0} = \left(\frac{\overline{D}_1}{\overline{D}_0} \right) IR. \tag{5.27}$$

If the average duration of an episode of the disease is the same whether someone is exposed or unexposed, that is, if $\overline{D}_1 = \overline{D}_0$, it follows from (5.27) that $POR = IR$. This shows that provided certain assumptions are satisfied, it is possible to estimate the crude incidence rate ratio using data from a prevalence study. In order for a causal interpretation to be possible, there must be temporality, and ideally also biologic plausibility (see Section 2.2). Identity (5.27) does not hold precisely for individual age groups, but in practice does so to an approximation.[1,10]

Example: National Comorbidity Survey Replication

Body mass index is defined to be weight (in kilograms) divided by the square of height (in meters). According to current standards, someone with a body mass index of 30 or more is considered to be obese. The National Comorbidity Survey Replication discussed in Section 3.1 examined the association between obesity at the time of interview (exposure) and the 1-year prevalence of major depression (disease), and found a prevalence odds ratio of 1.1.[11] The publication does not specify the temporal relationship of these two conditions, each of which can be a cause of the other, and so a causal interpretation is not possible. ◊

References

1. Miettinen O. Estimability and estimation in case-referent studies. *Am J Epidemiol* 1976;**103**(2):226–35.
2. Vandenbroucke JP, Pearce N. Incidence rates in dynamic populations. *Int J Epidemiol* 2012;**41**(5):1472–9.
3. Compressed Mortality File 1999–2016 at Centers for Disease Control and Prevention WONDER website https://wonder.cdc.gov. [Accessed May 2022].
4. Poole C. A history of the population attributable fraction and related measures. *Ann Epidemiol* 2015;**25**(3):147–54.
5. Kupper LL, Janis JM, Karmous A, et al. Statistical age-period-cohort analysis: a review and critique. *J Chron Dis* 1985;**38**(10):811–30.
6. Miettinen OS. *Theoretical epidemiology: principles of occurrence research in medicine*. New York: John Wiley & Sons; 1985.
7. Miettinen OS. Standardization of risk ratios. *Am J Epidemiol* 1972;**96**(6):383–8.

8. Holford TR. Understanding the effects of age, period, and cohort on incidence and mortality rates. *Annu Rev Public Health* 1991;**12**:425–57.
9. Freeman J, Hutchison GB. Prevalence, incidence and duration. *Am J Epidemiol* 1980;**112**(5):707–23.
10. Newman SC. Odds ratio estimation in a steady-state population. *J Clin Epidemiol* 1988;**41**(1):59–65.
11. Simon GE, Von Korff M, Saunders K, et al. Association between obesity and psychiatric disorders in the U.S. adult population. *Arch Gen Psychiatry* 2006;**63**(7):824–30.

CHAPTER

Effect Modification in Cohort Studies

6

The material in this chapter is framed in terms of closed cohort studies. The corresponding discussion of effect modification in open cohort studies is obtained by replacing risks with incidence rates in what follows and dropping any mention of odds ratios.

6.1 Effect Modification and Homogeneity

The concept of *effect modification* is most easily understood in the context of a closed cohort study of a disease with two exposures. Let us denote the disease by the variable D, and the exposures by the variables E and F. We assume that D and E are dichotomous with values $0, 1$ and, for the moment, that F is polychotomous with values $0, 1, \ldots, k$. We temporarily focus on E as the exposure of interest and treat F primarily as a stratifying variable.

Table 6.1 displays the risks corresponding to values of E and F. For example, $R_{E=1,F=i}$ is the risk of becoming a case for an exposed subject in stratum i of F, and $R_{E=1}$ is the overall risk of becoming a case for an exposed subject. In general, quantities that refer to a particular stratum are said to be *stratum-specific*, while those that are collapsed over strata are referred to as *crude*. For example, $R_{E=1,F=i}$ is a stratum-specific risk and $R_{E=1}$ is a crude risk.

The stratum-specific risk ratio, odds ratio and risk difference for the E-D association are

$$
\begin{aligned}
RR_{E|F=i} &= \frac{R_{E=1,F=i}}{R_{E=0,F=i}} \\
OR_{E|F=i} &= \frac{R_{E=1,F=i}/(1-R_{E=1,F=i})}{R_{E=0,F=i}/(1-R_{E=0,F=i})} \qquad (6.1) \\
RD_{E|F=i} &= R_{E=1,F=i} - R_{E=0,F=i}\,.
\end{aligned}
$$

The odds ratio has a more complicated structure than either the risk ratio or risk difference, and as a consequence does not share some of the properties of the risk ratio and risk difference to be discussed below. For that reason we will not consider the odds ratio further in this chapter.

Epidemiologic Methods. https://doi.org/10.1016/B978-0-44-318780-3.00012-9
Copyright © 2023 Elsevier Inc. All rights reserved.

Table 6.1 Risks for closed cohort study.

Stratum	$E=1$	$E=0$
$F=k$	$R_{E=1,F=k}$	$R_{E=0,F=k}$
$\vdots$	$\vdots$	$\vdots$
$F=i$	$R_{E=1,F=i}$	$R_{E=0,F=i}$
$\vdots$	$\vdots$	$\vdots$
$F=0$	$R_{E=1,F=0}$	$R_{E=0,F=0}$
Overall	$R_{E=1}$	$R_{E=0}$

The risk ratio for the E-D association is said to be *homogeneous* over F if the stratum-specific risk ratios are all equal across the strata of F, that is,

$$RR_{E|F=0} = RR_{E|F=1} = \cdots = RR_{E|F=k}, \tag{6.2}$$

and *heterogeneous* otherwise. When there is heterogeneity of the risk ratio, we say that F is an *effect modifier* of the risk ratio (for the E-D association). Similarly, the risk difference for the E-D association is said to be *homogeneous* over F if the stratum-specific risk differences are all equal across the strata of F, that is,

$$RD_{E|F=0} = RD_{E|F=1} = \cdots = RD_{E|F=k}, \tag{6.3}$$

and *heterogeneous* otherwise. When there is heterogeneity of the risk difference, we say that F is an *effect modifier* of the risk difference (for the E-D association). Observe that only the stratum-specific values of a measure of effect need to be considered when deciding whether effect modification is present; for that purpose, the crude value is irrelevant.

The presence of effect modification may have important practical implications. Suppose, for example, that E is a public health intervention directed at a disease D, and F is a demographic variable that is an effect modifier of the risk difference. In that situation, it may be prudent to focus the intervention on those demographic subgroups where it is likely to have the greatest impact as measured by the stratum-specific risk differences.

The above definitions of homogeneity have been presented in terms of strict equality. In the analysis of epidemiologic data, stratum-specific values of a measure of effect are never precisely equal. The question then becomes how close in value they need to be for the measure of effect to be considered homogeneous. This can be approached objectively by, for example, examining the extent to which stratum-specific confidence intervals overlap and using statistical tests of homogeneity. However, there is also a place for subjective opinion grounded in substantive knowledge of the exposure-disease association and practical considerations involving the purpose for which the epidemiologic study is intended. For example, when the sample size of the study is especially large, small differences in estimates of measures of effect might be statistically significant but of no practical importance. Since we are downplaying methods of statistical inference in this book, our assessment of homogeneity will be based on subjective impressions.

For the remainder of the chapter, we assume that F is dichotomous with values 0, 1. With this simplification, Table 6.2 takes the place of Table 6.1. We no longer view F as simply a stratifying variable, but rather as an exposure variable on an equal footing with E. To avoid redundancy, most definitions and related remarks will be made in terms of one exposure or the other, with the alternate definitions and remarks obtained by interchanging the roles of the two exposures. It needs to be emphasized that most of the results to follow apply only to the special situation considered here, where both exposures are dichotomous.

Table 6.2 Risks for closed cohort study.

	$F=1$	$F=0$
$E=1$	$R_{E=1,F=1}$	$R_{E=1,F=0}$
$E=0$	$R_{E=0,F=1}$	$R_{E=0,F=0}$

6.2 Effect Modification and Different Exposures

Having assumed that F is dichotomous, we have from (6.2) and (6.3) that the risk ratio for the E-D association is homogeneous over F if and only if

$$RR_{E|F=1} = RR_{E|F=0}, \tag{6.4}$$

and the risk difference for the E-D association is homogeneous over F if and only if

$$RD_{E|F=1} = RD_{E|F=0}. \tag{6.5}$$

It is demonstrated in Appendix E that:

F is an effect modifier of the risk ratio if and only if
E is an effect modifier of the risk ratio; (6.6)
and likewise for the risk difference.

This shows that effect modification has an inherent symmetry that applies separately to each of the measures of effect.

There is a way of characterizing homogeneity that offers additional insight. It is shown in Appendix E that the risk ratio for the E-D association is homogeneous over F if and only if

$$R_{E=1,F=1} = R_{E=0,F=0} \times RR_{E|F=0} \times RR_{F|E=0}, \tag{6.7}$$

and similarly, the risk difference for the E-D association is homogeneous over F if and only if

$$R_{E=1,F=1} = R_{E=0,F=0} + RD_{E|F=0} + RD_{F|E=0}. \tag{6.8}$$

We observe that (6.7) and (6.8) are symmetric in E and F.

By definition, $R_{E=0,F=0}$ is the risk of becoming a case when there is no exposure to E or F. Loosely speaking, (6.7) tells us that when the risk ratio for the E-D association is homogeneous over F, the risk due to exposure to both E and F equals the baseline risk, multiplied by a factor reflecting exposure to E, and then multiplied by a factor reflecting exposure to F. A similar interpretation can be given to (6.8).

Example: Hypothetical Closed Cohort Study

Table 6.3 gives the risks for a hypothetical closed cohort study.

Table 6.3 Risks for hypothetical closed cohort study.

	$F=1$	$F=0$
$E=1$	0.60	0.30
$E=0$	0.40	0.10

For the E-D association, with stratification by F:

$$RR_{E|F=1} = \frac{0.60}{0.40} = 1.5$$

$$RR_{E|F=0} = \frac{0.30}{0.10} = 3.0$$

$$RD_{E|F=1} = 0.60 - 0.40 = 0.20$$

$$RD_{E|F=0} = 0.30 - 0.10 = 0.20.$$

For the F-D association, with stratification by E:

$$RR_{F|E=1} = \frac{0.60}{0.30} = 2.0$$

$$RR_{F|E=0} = \frac{0.40}{0.10} = 4.0$$

$$RD_{F|E=1} = 0.60 - 0.30 = 0.30$$

$$RD_{F|E=0} = 0.40 - 0.10 = 0.30.$$

Both F and E are effect modifiers of the risk ratio, but neither is an effect modifier of the risk difference. This is consistent with (6.6). We note that (6.8) is satisfied $(0.60 = 0.10 + 0.20 + 0.30)$, but (6.7) is not $(0.60 \neq 0.10 \times 3.0 \times 4.0)$. ◊

Example: SARS Study

Table 6.4 is adapted from Table 7.9(a) and gives case fatality rates for the SARS study categorized by gender and age group.[1]

Table 6.4 Case fatality rates (percent) for SARS study.

Gender	Age group	
	45+	<45
Male	40.7	6.4
Female	30.1	2.8

Data from Reference 1.

Let E and F denote "gender" and "age group", respectively. For the gender-mortality association, with stratification by age group:

$$RR_{E|F=1} = \frac{40.7 \times 10^{-2}}{30.1 \times 10^{-2}} = 1.4$$

$$RR_{E|F=0} = \frac{6.4 \times 10^{-2}}{2.8 \times 10^{-2}} = 2.3$$

$$RD_{E|F=1} = (40.7 - 30.1) \times 10^{-2} = 10.6 \times 10^{-2}$$

$$RD_{E|F=0} = (6.4 - 2.8) \times 10^{-2} = 3.6 \times 10^{-2}.$$

For the age group-mortality association, with stratification by gender:

$$RR_{F|E=1} = \frac{40.7 \times 10^{-2}}{6.4 \times 10^{-2}} = 6.4$$

$$RR_{F|E=0} = \frac{30.1 \times 10^{-2}}{2.8 \times 10^{-2}} = 10.8$$

$$RD_{F|E=1} = (40.7 - 6.4) \times 10^{-2} = 34.3 \times 10^{-2}$$

$$RD_{F|E=0} = (30.1 - 2.8) \times 10^{-2} = 27.3 \times 10^{-2}.$$

Both age group and gender are effect modifiers of the risk ratio, and likewise for the risk difference. This is consistent with (6.6). The trends for risk ratios across strata are the opposite of trends for risk differences. This illustrates that patterns of effect modification depend on the choice of measure of effect. ◊

6.3 Effect Modification and Different Measures of Effect

In a closed cohort study, E might be associated with D only because E is associated with F, and F in turn is associated with D. In the absence of F, there would be no association between E and D. This can be expressed formally as $RR_{E|F=0} = 1$, or equivalently as $RD_{E|F=0} = 0$. For example, alcohol consumption is associated with lung cancer, not because it is a cause of this disease but because alcohol consumption is associated with cigarette smoking, and the latter is a cause of lung cancer.

The assertion that E is a risk factor for D independently of F can be formalized as $RR_{E|F=0} \neq 1$, or equivalently as $RD_{E|F=0} \neq 0$. Similarly, the assertion that F is a risk factor for D independently of E can be formalized as $RR_{F|E=0} \neq 1$, or equivalently as $RD_{F|E=0} \neq 0$. Consider the condition

$$RR_{E|F=0} \neq 1 \quad \text{and} \quad RR_{F|E=0} \neq 1\,, \tag{6.9}$$

which is equivalent to

$$RD_{E|F=0} \neq 0 \quad \text{and} \quad RD_{F|E=0} \neq 0\,. \tag{6.10}$$

It is shown in Appendix E that[2(p.71–83)]:

If (6.9) is true, or equivalently if (6.10) is true, then F is an effect modifier of the risk ratio or the risk difference, or both; and likewise for E. (6.11)

In other words, if E and F are risk factors for D independently of each other, then E and F are effect modifiers of the risk ratio or risk difference, or both. We note that for (6.9) not to be true, it is sufficient that either $RR_{E|F=0} = 1$ or $RR_{F|E=0} = 1$, but not necessarily both. A similar remark applies to (6.10).

Example: Down Syndrome Study

Table 6.5 is adapted from Table 7.10(a) and gives congenital anomaly rates categorized by birth order of the newborn and age group of the mother.[3]

Table 6.5 Congenital anomaly rates (per 100,000) for Down syndrome study.

Birth order	Age group	
	35+	<35
Third or greater	370.8	57.2
First or second	372.0	51.3

Data from Reference 2.

Let E and F denote "birth order of the newborn" and "age group of the mother", respectively. For the birth order-Down syndrome association, with stratification by age group:

$$RR_{E|F=1} = \frac{370.8 \times 10^{-5}}{372.0 \times 10^{-5}} = 1.00$$

$$RR_{E|F=0} = \frac{57.2 \times 10^{-5}}{51.3 \times 10^{-5}} = 1.12$$

$$RD_{E|F=1} = (370.8 - 372.0) \times 10^{-5} = -1.2 \times 10^{-5}$$

$$RD_{E|F=0} = (57.2 - 51.3) \times 10^{-5} = 5.9 \times 10^{-5}.$$

For the age group-Down syndrome association, with stratification by birth order:

$$RR_{F|E=1} = \frac{370.8 \times 10^{-5}}{57.2 \times 10^{-5}} = 6.48$$

$$RR_{F|E=0} = \frac{372.0 \times 10^{-5}}{51.3 \times 10^{-5}} = 7.25$$

$$RD_{F|E=1} = (370.8 - 57.2) \times 10^{-5} = 313.6 \times 10^{-5}$$

$$RD_{F|E=0} = (372.0 - 51.3) \times 10^{-5} = 320.7 \times 10^{-5}.$$

Taking a liberal view of what it means for stratum-specific measures of effect to be close enough in value to say that homogeneity is present, we find that age group is an effect modifier of the risk difference, but not an effect modifier of the risk ratio. We also find that birth order is not an effect modifier of either the risk ratio or risk difference. The latter observation does not conflict with (6.11) because (6.9) is not satisfied. With (6.9) not satisfied, we should also find that (6.10) is not satisfied. However, since the risks involved are so small, the corresponding risk differences are also very small, which makes it difficult to decide whether (6.10) holds for these data. ◊

6.4 Mathematical Models and Interaction

A distinction is sometimes made in the epidemiologic literature between *biologic interaction* and *statistical interaction*. The former refers to a type of dependency among biologic processes that takes place at the molecular, cellular or some higher functional level, while the latter is concerned with associations among study variables in a statistical model. As an example of biologic interaction, consider that the risk of infection from an infectious agent depends on the immune status of the individual: someone who is vaccinated will be protected, while someone who is immunocompromised will be susceptible. Epidemiologic methods are usually too unrefined to capture the intricacies of biologic processes. For this reason, we restrict our attention to statistical interaction, which for brevity we refer to simply as *interaction*.

Consider the additive model

$$R_{E,F} = a + bE + cF + dEF\,, \tag{6.12}$$

where $R_{E,F}$, the risk of becoming a case, is taken to be a function of the dichotomous variables E and F, and where a, b, c and d are constants to be estimated from study data. The interpretation of model (6.12) is that for given values of E and F, the corresponding risk equals a baseline risk (given by a), plus the direct effects of E and F operating separately (given by bE and cF, respectively), plus an interaction effect due to E and F operating jointly (given by dEF). Table 6.6 gives the risks based on model (6.12) for values of E and F.

Table 6.6 Risks based on model (6.12).

	$F = 1$	$F = 0$
$E = 1$	$R_{E=1,F=1} = a + b + c + d$	$R_{E=1,F=0} = a + b$
$E = 0$	$R_{E=0,F=1} = a + c$	$R_{E=0,F=0} = a$

For the E-D association, with stratification by F:

$$\begin{aligned} RD_{E|F=1} &= (a + b + c + d) - (a + c) = b + d \\ RD_{E|F=0} &= (a + b) - a = b\,, \end{aligned} \tag{6.13}$$

which shows that F is an effect modifier of the risk difference if and only if $d \neq 0$. That is, F is an effect modifier of the risk difference if and only if model (6.12) has an interaction term. For the F-D association, with stratification by E:

$$\begin{aligned} RD_{F|E=1} &= (a + b + c + d) - (a + b) = c + d \\ RD_{F|E=0} &= (a + c) - a = c\,, \end{aligned} \tag{6.14}$$

from which we see that E is an effect modifier of the risk difference if and only if $d \neq 0$. The symmetry exhibited by model (6.12) with respect to effect modification by E and F is reminiscent of (6.6).

Suppose there is no interaction, that is, suppose $d = 0$, in which case $R_{E=1,F=1} = a + b + c$. We have from Table 6.6 that $a = R_{E=0,F=0}$, and from (6.13) and (6.14) that $b = RD_{E|F=0}$ and $c = RD_{F|E=0}$. Combining these identities yields (6.8).

An advantage of using model (6.12) is that it captures the essential features of effect modification and can be expanded to accommodate additional exposures and more complex relationships among them. A disadvantage is that like all mathematical models, model (6.12) is only as valid as the assumptions underlying it. In particular, model (6.12) assumes that exposures act additively, something that may be difficult to justify.

6.5 Qualitative Effect Modification

Suppose F is an effect modifier of a given measure of effect. We say that F is a *qualitative effect modifier* if one of the stratum-specific measures of effect is greater than its null value, and the other is less than its null value.[4, 5(p.867–871)] Expressed more formally, F is a qualitative effect modifier of the risk ratio if and only if

$$RR_{E|F=0} < 1 < RR_{E|F=1} \quad \text{or} \quad RR_{E|F=0} > 1 > RR_{E|F=1} .$$

Likewise, F is a qualitative effect modifier of the risk difference if and only if

$$RD_{E|F=0} < 0 < RD_{E|F=1} \quad \text{or} \quad RD_{E|F=0} > 0 > RD_{E|F=1} .$$

If F is an effect modifier but not a qualitative effect modifier, it is said to be a *quantitative effect modifier*.

It was remarked above that the presence of effect modification may have important practical implications. This is especially true in the medical setting when there is qualitative effect modification. Suppose, for example, that E is a treatment for a disease D, and F is a clinical characteristic that is a qualitative effect modifier of the risk ratio. In that situation, the treatment will be beneficial in one of the clinical subgroups and harmful in the other.

It is shown in Appendix E that:

$$\begin{gathered}F \text{ is a qualitative effect modifier of the risk ratio if and only if} \\ F \text{ is a qualitative effect modifier of the risk difference;} \\ \text{and likewise for } E.\end{gathered} \tag{6.15}$$

It is also shown in Appendix E that when F is a qualitative effect modifier of the risk ratio or the risk difference (hence a qualitative effect modifier of both), the stratum-specific values of the risk ratio and risk difference exhibit the same trend over the strata of F. That is, both measures of effect either increase across strata, or they both decrease across strata. As is demonstrated by the SARS example, this feature of qualitative effect modification does not carry over to quantitative effect modification. These observations show that when there is qualitative effect modification, we get essentially the same assessment of effect modification whether we choose the risk ratio or risk difference as the measure of effect.

The next example illustrates a type of interaction seen in multivariate models called *crossover interaction*. It can be thought of as "qualitative effect modification" where some or all of the exposure variables are continuous.

Example: Birth Weight Paradox

The proportion of newborns dying in the first year of life (that is, the 1-year mortality risk for newborns) is traditionally called the *infant mortality rate*. It is well-established in the pediatric literature that maternal smoking during pregnancy and

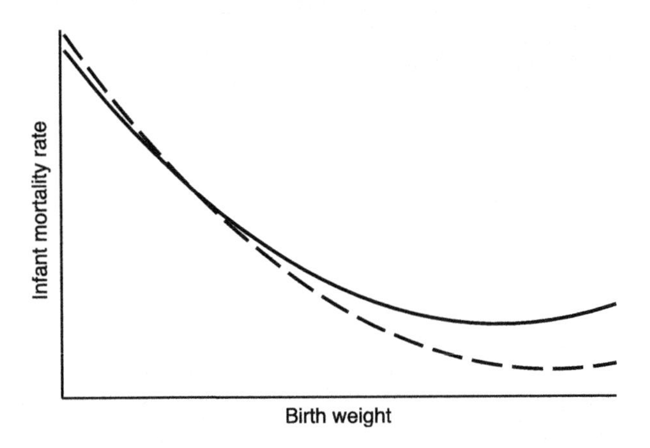

FIGURE 6.1

Schematic diagram depicting infant mortality rate as a function of birth weight, with stratification by smoking status of mother during pregnancy; vertical axis uses logarithmic scale (smoker: *solid*; nonsmoker: *dash*).

low birth weight are associated with increased infant mortality.[6] It is therefore reasonable to expect that a graph of the infant mortality rate as a function of birth weight, with stratification by smoking status of the mother (smoked during pregnancy, did not smoke during pregnancy), would consist of two downward sloping curves, with the curve for mothers who smoked standing above the curve for mothers who did not. It turns out that this is not the reality.

Fig. 6.1 is a schematic diagram of the graph that is usually seen in practice, where the curves cross over at a point roughly corresponding to a birth weight of 2,500 grams, the generally-accepted threshold for defining the low birth weight category.[6–8] The interpretation is that in normal birth weight newborns, the infant mortality rate is higher in the offspring of mothers who smoked than in the offspring of mothers who did not. But in low birth weight newborns, the infant mortality rate is lower in the offspring of mothers who smoked than in the offspring of mothers who did not. Expressed in terms of a multivariate model, there is a crossover interaction involving maternal smoking and birth weight. The implication seems to be that maternal smoking during pregnancy somehow protects low birth weight newborns from premature mortality, a counterintuitive finding called the *birth weight paradox*.[6–8] A possible explanation for this phenomenon is provided in Section 8.1. ◊

The next example also considers the birth weight paradox, but this time with birth weight as a dichotomous variable.

Example: Birth Weight Paradox

A closed cohort study of the association between maternal smoking during pregnancy and infant mortality was conducted in which all singleton live births in the

United States in 1997 were followed for one year, with death from any cause as the outcome of interest.[9] Table 6.7 gives the infant mortality rates categorized by smoking status of the mother during pregnancy (smoker, nonsmoker) and birth weight of the newborn as a dichotomous variable (low, normal). Consistent with Fig. 6.1, in normal birth weight newborns the infant mortality rate is higher in the offspring of mothers who smoked than in the offspring of mothers who did not, but in low birth weight newborns the opposite is true. This again illustrates the birth weight paradox.

Table 6.7 Infant mortality rates (per 10,000) for infant mortality study.

Smoking status	Birth weight	
	Low	Normal
Smoker	542.8	48.9
Nonsmoker	684.4	23.8

Data from Reference 9.

Let E and F denote "smoking status of the mother" and "birth weight of the newborn", respectively. For the smoking status-mortality association, with stratification by birth weight:

$$RR_{E|F=1} = \frac{542.8 \times 10^{-4}}{684.4 \times 10^{-4}} = 0.79$$

$$RR_{E|F=0} = \frac{48.9 \times 10^{-4}}{23.8 \times 10^{-4}} = 2.05$$

$$RD_{E|F=1} = (542.8 - 684.4) \times 10^{-4} = -141.6 \times 10^{-4}$$

$$RD_{E|F=0} = (48.9 - 23.8) \times 10^{-4} = 25.1 \times 10^{-4}.$$

For the birth weight-mortality association, with stratification by smoking status:

$$RR_{F|E=1} = \frac{542.8 \times 10^{-4}}{48.9 \times 10^{-4}} = 11.1$$

$$RR_{F|E=0} = \frac{684.4 \times 10^{-4}}{23.8 \times 10^{-4}} = 28.8$$

$$RD_{F|E=1} = (542.8 - 48.9) \times 10^{-4} = 493.9 \times 10^{-4}$$

$$RD_{F|E=0} = (684.4 - 23.8) \times 10^{-4} = 660.6 \times 10^{-4}.$$

Birth weight is a qualitative effect modifier of the risk ratio and risk difference, and the trend over the strata of birth weight is the same for both measures of effect.

This is consistent with (6.15) and subsequent remarks. On the other hand, smoking status is a quantitative effect modifier of the risk ratio and risk difference. ◊

References

1. Karlberg J, Chong DSY, Lai WYY. Do men have a higher case fatality rate of severe acute respiratory syndrome that women do? *Am J Epidemiol* 2004;**159**(3):229–31.
2. Greenland S, Lash TL, Rothman KJ. Concepts of interaction. In: Rothman KJ, Greenland S, Lash TL, editors. *Modern epidemiology*. 3rd ed. Philadelphia: Lippincott Williams & Wilkins; 2008.
3. Stark CR, Mantel N. Effects of maternal age and birth order on the risk of mongolism and leukemia. *J Natl Cancer Inst* 1966;**37**(5):687–98.
4. VanderWeele TJ, Knol MJ. A tutorial on interaction. *Epidemiol Methods* 2014;**3**(1):33–72.
5. Peto R. Statistical aspects of cancer trials. In: Halnan KE, editor. *Treatment of cancer*. London: Chapman and Hall; 1982.
6. Wilcox AJ. On the importance—and the unimportance—of birthweight. *Int J Epidemiol* 2001;**30**(6):1233–41.
7. Hernández-Díaz S, Schisterman EF, Hernán MA. The birth weight "paradox" uncovered? *Am J Epidemiol* 2006;**164**(11):1115–20.
8. Wilcox AJ. Birth weight and perinatal mortality: the effect of maternal smoking. *Am J Epidemiol* 1993;**137**(10):1098–104.
9. VanderWeele TJ, Mumford SL, Schisterman EF. Conditioning on intermediates in perinatal epidemiology. *Epidemiology* 2012;**23**(1):1–9.

CHAPTER

7 Standardization in Cohort Studies

It was demonstrated in Chapter 5 that the comparison of incidence in two populations can be distorted when the person-years distributions of those at risk in the populations differ. The same type of problem can arise when comparing incidence in two cohorts. Our aim in what follows is to adapt the methods of indirect standardization introduced in Section 5.3 for stationary populations to the setting of cohort studies. It was pointed out there that indirect standardization is nothing more than a special case of direct standardization. Through historical precedent and long usage, the indirect-direct distinction is well entrenched in demography. However, there is no need to carry it into the discussion of cohort studies. Accordingly, we will speak of "standardization" in cohort studies rather than "indirect standardization". The definitions and notation will make it clear that we are discussing the cohort study counterpart of what was referred to as indirect standardization in stationary populations.

7.1 Closed Cohort Studies

Let us begin by introducing some notation. We continue with the closed cohort study described at the beginning of Chapter 6, but now take E to be a dichotomous exposure of interest and treat F exclusively as a polychotomous stratifying variable. This change in perspective allows us to significantly simplify notation. Table 7.1 presents the earlier notation alongside its revised counterpart. Table 7.2 gives the data and risks for stratum i in the revised notation using the classical 2×2 layout. Table 7.3 displays crude and stratum-specific data and risks using an alternative layout. For typographical convenience, the columns for noncases in Table 7.3 have been suppressed.

The stratum-specific risk ratio, odds ratio and risk difference, previously defined in (6.1), can now be expressed simply as

$$RR_i = \frac{R_{1i}}{R_{0i}}$$

$$OR_i = \frac{R_{1i}/(1 - R_{1i})}{R_{0i}/(1 - R_{0i})} \tag{7.1}$$

$$RD_i = R_{1i} - R_{0i} \,.$$

Epidemiologic Methods. https://doi.org/10.1016/B978-0-44-318780-3.00013-0
Copyright © 2023 Elsevier Inc. All rights reserved.

Table 7.1 Notation from Chapter 6 and revised notation.

Chapter 6	Revised
$R_{E=1,F=i}$	R_{1i}
$R_{E=0,F=i}$	R_{0i}
$R_{E=1}$	R_1
$R_{E=0}$	R_0
$RR_{E\|F=i}$	RR_i
$OR_{E\|F=i}$	OR_i
$RD_{E\|F=i}$	RD_i

Table 7.2 Stratum-specific data and risks for closed cohort study.

Exposure	Cases	Noncases	Persons	Risk
$E=1$	a_{1i}	b_{1i}	n_{1i}	R_{1i}
$E=0$	a_{0i}	b_{0i}	n_{0i}	R_{0i}

Table 7.3 Crude and stratum-specific data and risks for closed cohort study.

Stratum	$E=1$			$E=0$		
	Cases	Persons	Risk	Cases	Persons	Risk
$F=k$	a_{1k}	n_{1k}	R_{1k}	a_{0k}	n_{0k}	R_{0k}
$\vdots$	$\vdots$	$\vdots$	$\vdots$	$\vdots$	$\vdots$	$\vdots$
$F=i$	a_{1i}	n_{1i}	R_{1i}	a_{0i}	n_{0i}	R_{0i}
$\vdots$	$\vdots$	$\vdots$	$\vdots$	$\vdots$	$\vdots$	$\vdots$
$F=0$	a_{10}	n_{10}	R_{10}	a_{00}	n_{00}	R_{00}
Overall	a_1	n_1	R_1	a_0	n_0	R_0

The distribution of F in the exposed cohort at the start of follow-up is

$$\frac{n_{10}}{n_1}, \ldots, \frac{n_{1i}}{n_1}, \ldots, \frac{n_{1k}}{n_1},$$

and in the unexposed cohort is

$$\frac{n_{00}}{n_0}, \ldots, \frac{n_{0i}}{n_0}, \ldots, \frac{n_{0k}}{n_0}.$$

Standardized Measures of Effect

It can be shown that R_1 and R_0 can be expressed as

$$R_1 = \sum_{i=0}^{k} \left(\frac{n_{1i}}{n_1}\right) R_{1i}$$
$$R_0 = \sum_{i=0}^{k} \left(\frac{n_{0i}}{n_0}\right) R_{0i}\,. \tag{7.2}$$

Similar to the situation with (5.11), we see from (7.2) that differing distributions of F in the exposed and unexposed cohorts at the start of follow-up can distort the comparison of crude risks. Our response to the problem is analogous to the approach taken in Chapter 5. Following (5.16), the *standardized risk* in the unexposed cohort, with the exposed cohort as the "standard cohort", is defined to be

$$R_{0(1)} = \sum_{i=0}^{k} \left(\frac{n_{1i}}{n_1}\right) R_{0i}\,. \tag{7.3}$$

We interpret $R_{0(1)}$ as the crude risk that the unexposed cohort would have experienced if it had had the distribution of F in the exposed cohort at the start of follow-up, but retained its stratum-specific risks. Unlike the corresponding situation with the indirectly standardized incidence rate in a stationary population, the stratum-specific risks do not affect the distribution of F at the start of follow-up. So our intuitive interpretation of $R_{0(1)}$ is meaningful.

The *standardized risk ratio*, *standardized odds ratio* and *standardized risk difference* are defined to be

$$sRR = \frac{R_1}{R_{0(1)}}$$
$$sOR = \frac{R_1/(1-R_1)}{R_{0(1)}/(1-R_{0(1)})} \tag{7.4}$$
$$sRD = R_1 - R_{0(1)}\,,$$

respectively. That is, the standardized measures of effect are obtained from the crude measures of effect in (4.3) by replacing R_0 with $R_{0(1)}$. It is shown in Appendix F that sRR is a weighted average of stratum-specific risk ratios, and sRD is a weighted average of stratum-specific risk differences. However, sOR is not in general a weighted average of stratum-specific odds ratios.[1(p.62–63)]

The *observed number* of incident cases in the exposed cohort is a_1. Following (5.17), let us define the *expected number* of incident cases in the exposed cohort to be[2,3]

$$a_{0(1)} = \sum_{i=0}^{k} R_{0i} n_{1i} = R_{0(1)} n_1 , \tag{7.5}$$

where the second equality follows from (7.3). We interpret $a_{0(1)}$ as the number of incident cases that would have occurred in the exposed cohort if it had had the stratum-specific risks of the unexposed cohort, but retained the distribution of F at the start of follow-up. For the same reasons given in connection with the standardized risk, this intuitive interpretation of $a_{0(1)}$ is justified. Corresponding to (5.18), we find that the standardized risk ratio can be expressed as[2,3]

$$sRR = \frac{a_1}{a_{0(1)}} . \tag{7.6}$$

Mantel-Haenszel Measures of Effect

There are alternatives to the standardized measures of effect that are designed specifically for the situation where the measures of effect are homogeneous. When there is homogeneity, the most widely-used of the classical estimators are those based on the approach of Mantel and Haenszel.[4] The *Mantel-Haenszel risk ratio*, *Mantel-Haenszel odds ratio* and *Mantel-Haenszel risk difference* are defined to be[1(p.132,148,155–156), 5(p.179)]

$$\begin{aligned} RR_{MH} &= \sum_{i=0}^{k} \frac{a_{1i} n_{0i}}{n_{0i} + n_{1i}} \Bigg/ \sum_{i=0}^{k} \frac{a_{0i} n_{1i}}{n_{0i} + n_{1i}} \\ OR_{MH} &= \sum_{i=0}^{k} \frac{a_{1i} b_{0i}}{n_{0i} + n_{1i}} \Bigg/ \sum_{i=0}^{k} \frac{a_{0i} b_{1i}}{n_{0i} + n_{1i}} \\ RD_{MH} &= \sum_{i=0}^{k} \frac{a_{1i} n_{0i} - a_{0i} n_{1i}}{n_{0i} + n_{1i}} \Bigg/ \sum_{i=0}^{k} \frac{n_{0i} n_{1i}}{n_{0i} + n_{1i}} , \end{aligned} \tag{7.7}$$

respectively. It is shown in Appendix F that RR_{MH} is a weighted average of stratum-specific risk ratios, OR_{MH} is a weighted average of stratum-specific odds ratios, and RD_{MH} is a weighted average of stratum-specific risk differences.

When the risk ratio (resp. odds ratio, risk difference) is homogeneous, we denote the common stratum-specific value by hRR (resp. hOR, hRD). It follows from item 3 of Section 2.4 and the above remarks that:

1. If the risk ratio is homogeneous, then

$$sRR = hRR = RR_{MH} . \tag{7.8}$$

2. If the risk difference is homogeneous, then

$$sRD = hRD = RD_{MH} . \tag{7.9}$$

3. If the odds ratio is homogeneous, then

$$OR_{\mathrm{MH}} = hOR \quad \text{but maybe} \quad sOR \neq hOR\,. \tag{7.10}$$

It is shown in Appendix F that if the odds ratio is homogeneous and the risk of becoming a case is "small", we have the approximation $sOR \approx hOR$. Then (7.10) can be replaced with

$$sOR \approx hOR = OR_{\mathrm{MH}}\,. \tag{7.11}$$

A major difference between the standardized and Mantel-Haenszel measures of effect is that the weights for the latter reflect the amount of statistical "information" in each stratum, with strata having relatively more information receiving greater weight. For this reason, Mantel-Haenszel measures of effect are statistically "efficient", meaning that, all other things being equal, their estimates have relatively small variances. However, Mantel-Haenszel measures of effect have the drawback that, strictly speaking, they should only be used when there is homogeneity. In practice, the interpretation of what constitutes homogeneity tends to be flexible.

Closed Cohort Studies #1–3

To illustrate some of the above theory, we consider three hypothetical closed cohort studies, referred to as closed cohort studies #1, #2 and #3. For each of the studies, E is the exposure and F is the stratifying variable. Tables 7.6(a), 7.7(a) and 7.8(a) give the stratum-specific data and risks. Tables 7.6(b), 7.7(b) and 7.8(b) give the distribution of F at the start of follow-up. Tables 7.6(c), 7.7(c) and 7.8(c) give the crude, stratum-specific, standardized and Mantel-Haenszel risk ratios, odds ratios and risk differences. As an example of the computations involved in Tables 7.6(c), 7.7(c) and 7.8(c), we have for closed cohort study #1 that

$$R_1 = \frac{170}{300} = 0.567$$

and

$$R_{0(1)} = \left(\frac{200}{300}\right)\left(\frac{20}{200}\right) + \left(\frac{100}{300}\right)\left(\frac{60}{100}\right) = 0.267\,,$$

hence

$$sRR = \frac{0.567}{0.267} = 2.13\,.$$

For closed cohort study #1, we see from Table 7.6(b) that the distribution of F at the start of follow-up is the same in the exposed and unexposed cohorts, and from Table 7.6(c) that the crude measures of effect equal their standardized counterparts. In fact, the second finding is a consequence of the first. Observe that in Table 7.6(c) the odds ratio and risk difference are homogeneous. Consistent with (7.9) and (7.10),

$$sRD = 3.00 = RD_{MH}$$

and

$$OR_{MH} = 6.00 \quad \text{and} \quad sOR = 3.60\,.$$

For closed cohort study #2, we have from Table 7.7(b) that the distribution of F at the start of follow-up differs in the exposed and unexposed cohorts, and from Table 7.7(c) that the crude measures of effect differ substantially from their standardized counterparts. It is noteworthy that the crude measures of effect reflect a negative E-D association, while the standardized and Mantel-Haenszel measures of effect indicate a positive one. Observe that in Table 7.7(c), the risk ratio is homogeneous. Consistent with (7.8),

$$sRR = 1.50 = RR_{MH}\,.$$

For closed cohort study #3, we see from Table 7.8(b) that the distribution of F at the start of follow-up is different in the exposed and unexposed cohorts, and from Table 7.8(c) that the crude measures of effect equal their standardized counterparts. This shows that a difference in the distribution of F at the start of follow-up in the exposed and unexposed cohorts is not sufficient to ensure that crude measures of effect differ from their standardized counterparts. Note that in Table 7.8(c), the risk difference is homogeneous. Consistent with (7.9),

$$sRD = 2.0 = RD_{MH}\,.$$

SARS and Down Syndrome Studies

We close this section with a few observations on the SARS and Down syndrome studies discussed in Section 4.1.

Example: SARS Study

For the SARS study, we consider stratification by age group. Table 7.9(a) gives the age-specific data and case fatality rates.[6] Table 7.9(b) gives the age distribution at the start of follow-up. Table 7.9(c) gives the crude, age-specific, standardized and Mantel-Haenszel risk ratios, odds ratios and risk differences. Case fatality rates increase with increasing age group for both males and females. The age distributions at the start of follow-up differ somewhat for males and females. There is effect modification of each of the measures of effect. ◊

Example: Down Syndrome Study

For the Down syndrome study, we consider stratification by maternal age group.[7] Table 7.10(a) gives the age-specific data and congenital anomaly rates. Table 7.10(b) gives the age distribution at the start of follow-up. Table 7.10(c) gives the crude, age-specific, standardized and Mantel-Haenszel risk ratios, odds ratios and risk differences. Congenital anomaly rates increase markedly with increasing maternal age group for both birth order categories. The age distributions at the start of follow-up differ substantially across birth order categories. Adopting a liberal interpretation of what it means for stratum-specific values to be close in value, the risk ratio and odds ratio are homogeneous, but the risk difference is not. ◊

7.2 Open Cohort Studies

The following discussion of standardization and Mantel-Haenszel methods for open cohort studies closely parallels that given above for closed cohort studies.

Table 7.4 gives the data and incidence rates for stratum i of an open cohort study using the classical 2×2 layout, and Table 7.5 gives the crude and stratum-specific data and incidence rates using an alternative layout.

Table 7.4 Stratum-specific data and incidence rates for open cohort study.

Exposure	Cases	Person-time	Rate
$E = 1$	a_{1i}	y_{1i}	I_{1i}
$E = 0$	a_{0i}	y_{0i}	I_{0i}

Table 7.5 Stratum-specific data and incidence rates for open cohort study.

Stratum	$E = 1$			$E = 0$		
	Cases	Person-time	Rate	Cases	Person-time	Rate
$F = k$	a_{1k}	y_{1k}	I_{1k}	a_{0k}	y_{0k}	I_{0k}
⋮	⋮	⋮	⋮	⋮	⋮	⋮
$F = i$	a_{1i}	y_{1i}	I_{1i}	a_{0i}	y_{0i}	I_{0i}
⋮	⋮	⋮	⋮	⋮	⋮	⋮
$F = 0$	a_{10}	y_{10}	I_{10}	a_{00}	y_{00}	I_{00}
Overall	a_1	y_1	I_1	a_0	y_0	I_0

The stratum-specific incidence rate ratio and incidence rate difference are

$$IR_i = \frac{I_{1i}}{I_{0i}}$$

$$ID_i = I_{1i} - I_{0i} ,$$

respectively. The distribution of person-years over the strata of F in the exposed cohort is

$$\frac{y_{10}}{y_1}, \ldots, \frac{y_{1i}}{y_1}, \ldots, \frac{y_{1k}}{y_1},$$

and in the unexposed cohort is

$$\frac{y_{00}}{y_0}, \ldots, \frac{y_{0i}}{y_0}, \ldots, \frac{y_{0k}}{y_0}.$$

Standardized and Mantel-Haenszel Measures of Effect

It can be shown that I_1 and I_0 can be expressed as

$$I_1 = \sum_{i=0}^{k} \left(\frac{y_{1i}}{y_1} \right) I_{1i}$$

$$I_0 = \sum_{i=0}^{k} \left(\frac{y_{0i}}{y_0} \right) I_{0i} .$$

Following (5.16) and (7.3), the *standardized incidence rate* in the unexposed cohort, with the exposed cohort as the "standard cohort", is defined to be

$$I_{0(1)} = \sum_{i=0}^{k} \left(\frac{y_{1i}}{y_1} \right) I_{0i} .$$

We would like to interpret $I_{0(1)}$ as the crude incidence rate that the unexposed cohort would have experienced if it had had the person-years distribution of the exposed cohort, but retained its stratum-specific incidence rates. However, we run into the same problem that was encountered with the interpretation of the standardized incidence rate in (5.12), namely, that the person-years distribution of the unexposed cohort is determined in part by its stratum-specific incidence rates.

The *standardized incidence rate ratio* and *standardized incidence rate difference* are defined to be

$$sIR = \frac{I_1}{I_{0(1)}} \qquad (7.12)$$

$$sID = I_1 - I_{0(1)} ,$$

respectively. The *Mantel-Haenszel incidence rate ratio* and *Mantel-Haenszel incidence rate difference* are defined to be[1(p.223–224), 5(p.184)]

$$IR_{\text{MH}} = \sum_{i=0}^{k} \frac{a_{1i}\, y_{0i}}{y_{0i} + y_{1i}} \Bigg/ \sum_{i=0}^{k} \frac{a_{0i}\, y_{1i}}{y_{0i} + y_{1i}}$$

$$ID_{\text{MH}} = \sum_{i=0}^{k} \frac{a_{1i}\, y_{0i} - a_{0i}\, y_{1i}}{y_{0i} + y_{1i}} \Bigg/ \sum_{i=0}^{k} \frac{y_{0i}\, y_{1i}}{y_{0i} + y_{1i}}, \tag{7.13}$$

respectively. It is shown in Appendix F that sIR and IR_{MH} are weighted averages of stratum-specific incidence rate ratios, and sID and ID_{MH} are weighted averages of stratum-specific incidence rate differences.

Example: British Doctors' Study

For the British Doctors' Study discussed in Section 4.2, we consider stratification by age group.[8(p.205-268)] Tables 7.11(a) and 7.11(b) give the age-specific data, mortality rates and mortality rate ratios, and Table 7.11(c) gives the crude, standardized and Mantel-Haenszel mortality rate ratios and mortality rate differences. The standardized and Mantel-Haenszel rate ratios and rate differences provide evidence of increased coronary heart disease mortality in smokers compared to nonsmokers, after accounting for differences in person-years distribution.

Methods corresponding to direct standardization in geographically-defined populations can be used to analyze data from a large open cohort study such as the British Doctors' Study. Let us take the male population of England and Wales in 1956 to be the standard population (population $*$). The directly standardized mortality rates in smokers and nonsmokers are $I_{1(*)} = 56.81 \times 10^{-4}$ per year and $I_{0(*)} = 42.42 \times 10^{-4}$ per year, respectively. The directly standardized mortality ratio is

$$dMR = \frac{56.81 \times 10^{-4}}{42.42 \times 10^{-4}} = 1.34,$$

which is close in value to the (indirectly) standardized and Mantel-Haenszel mortality rate ratios in Table 7.11(c). ◊

Table 7.6(a) Data and risks for closed cohort study #1.

Stratum	$E=1$			$E=0$		
	Cases	Persons	Risk	Cases	Persons	Risk
$F=1$	90	100	0.90	60	100	0.60
$F=0$	80	200	0.40	20	200	0.10
Overall	170	300	0.57	80	300	0.27

Table 7.6(b) Distribution of F at start of follow-up (percent) for closed cohort study #1.

Stratum	$E=1$	$E=0$
$F=1$	33.3	33.3
$F=0$	66.7	66.7

Table 7.6(c) Measures of effect for closed cohort study #1: E is exposure and F is stratification variable.

Measure of effect	Crude	Stratum-specific		Stand.	M-H
		$F=1$	$F=0$		
Risk ratio	2.13	1.50	4.00	2.13	2.13
Odds ratio	3.60	6.00	6.00	3.60	6.00
Risk difference[a]	3.00	3.00	3.00	3.00	3.00

[a] *per 10.*

Table 7.7(a) Data and risks for closed cohort study #2.

Stratum	$E=1$			$E=0$		
	Cases	Persons	Risk	Cases	Persons	Risk
$F=1$	90	100	0.90	120	200	0.60
$F=0$	30	200	0.15	10	100	0.10
Overall	120	300	0.40	130	300	0.43

Table 7.7(b) Distribution of F at start of follow-up (percent) for closed cohort study #2.

Stratum	$E=1$	$E=0$
$F=1$	33.3	66.7
$F=0$	66.7	33.3

Table 7.7(c) Measures of effect for closed cohort study #2: E is exposure and F is stratification variable.

Measure of effect	Crude	Stratum-specific		Stand.	M-H
		$F=1$	$F=0$		
Risk ratio	0.92	1.50	1.50	1.50	1.50
Odds ratio	0.87	6.00	1.59	1.83	3.41
Risk difference[a]	−0.33	0.30	0.50	1.33	1.75

[a] *per 10.*

Table 7.8(a) Data and risks for closed cohort study #3.

Stratum	$E=1$			$E=0$		
	Cases	Persons	Risk	Cases	Persons	Risk
$F=2$	50	100	0.50	90	300	0.30
$F=1$	90	300	0.30	20	200	0.10
$F=0$	140	200	0.70	50	100	0.50
Overall	280	600	0.47	160	600	0.27

Table 7.8(b) Distribution of F at start of follow-up (percent) for closed cohort study #3.

Stratum	$E=1$	$E=0$
$F=2$	16.7	50.0
$F=1$	50.0	33.3
$F=0$	33.3	16.7

Table 7.8(c) Measures of effect for closed cohort study #3: E is exposure and F is stratification variable.

Measure of effect	Crude	Stratum-specific			Stand.	M-H
		$F=2$	$F=1$	$F=0$		
Risk ratio	1.75	1.67	3.00	1.40	1.75	1.77
Odds ratio	2.41	2.33	3.86	2.33	2.41	2.77
Risk difference[a]	2.00	2.00	2.00	2.00	2.00	2.00

[a] *per 10.*

Table 7.9(a) Data and case fatality rates for SARS study.

Age group	Male			Female		
	Deaths	Persons	Rate[a]	Deaths	Persons	Rate[a]
75+	66	102	64.7	49	77	63.6
45–74	77	249	30.9	63	295	21.4
<45	27	425	6.4	17	607	2.8
Overall	170	776	21.9	129	979	13.2

[a] *per 100.*
Data from Reference 6.

Table 7.9(b) Age distribution at start of follow-up (percent) for SARS study.

Age group	Male	Female
75+	13.1	7.9
45–74	32.1	30.1
<45	54.8	62.0

Data from Reference 6.

Table 7.9(c) Measures of effect for SARS study: gender is exposure and age group is stratifying variable.

Measure of effect	Crude	Age group			Stand.	M-H
		75+	45–74	<45		
Risk ratio	1.66	1.02	1.45	2.27	1.31	1.35
Odds ratio	1.85	1.05	1.65	2.35	1.39	1.61
Risk difference[a]	8.73	1.07	9.57	3.55	5.16	5.19

[a] *per 100.*
Data from Reference 6.

Table 7.10(a) Data and congenital anomaly rates for Down syndrome study.

Age group	Birth order 3+			Birth order <3		
	Cases	Live births	Rate[a]	Cases	Live births	Rate[a]
35+	928	250,293	370.8	183	49,198	372.0
<35	618	1,079,807	57.2	694	1,352,431	51.3
Overall	1,546	1,330,100	116.2	877	1,401,629	62.6

[a] *per 100,000.*
Data from Reference 7.

Table 7.10(b) Age distribution of mothers (percent) for Down syndrome study.

Age group	Birth order	
	3+	<3
35+	18.8	3.5
<35	81.2	96.5

Data from Reference 7.

Table 7.10(c) Measures of effect for Down syndrome study: birth order is exposure and age group is stratifying variable.

Measure of effect	Crude	Age group		Stand.	M-H
		35+	<35		
Risk ratio	1.86	1.00	1.12	1.04	1.08
Odds ratio	1.86	1.00	1.12	1.04	1.08
Risk difference[a]	53.66	−1.20	5.92	4.58	5.46

[a] *per 100,000.*
Data from Reference 7.

Table 7.11(a) Data for British Doctors' Study.

Age group	Smoker		Nonsmoker	
	Deaths	Person-years	Deaths	Person-years
35–44	32	52,407	2	18,790
45–54	104	43,248	12	10,673
55–64	206	28,612	28	5,710
65–74	186	12,663	28	2,585
75–84	102	5,317	31	1,462
85+	30	835	11	340
Overall	660	143,082	112	39,560

Data from Reference 8.

Table 7.11(b) Coronary heart disease mortality rates for British Doctors' Study.

Age group	Mortality rate[a]		Rate ratio
	Smoker	Nonsmoker	
35–44	6.1	1.1	5.74
45–54	24.0	11.2	2.14
55–64	72.0	49.0	1.47
65–74	146.9	108.3	1.36
75–84	191.8	212.0	0.90
85+	359.3	323.5	1.11
Overall	46.1	28.3	1.63

[a] *per 10,000 per year.*
Data from Reference 8.

Table 7.11(c) Measures of effect for British Doctors' Study: smoking is exposure and age group is stratifying variable.

Measure of effect	Crude	Stand.	M-H
Rate ratio	1.63	1.40	1.40
Rate difference[a]	17.82	13.18	11.63

[a] *per 10,000 per year.*
Data from Reference 8.

References

1. Newman SC. *Biostatistical methods in epidemiology.* New York: Wiley; 2001.
2. Miettinen OS. Components of the crude risk ratio. *Am J Epidemiol* 1972;**96**(2):168–72.
3. Miettinen OS. Standardization of risk ratios. *Am J Epidemiol* 1972;**96**(6):383–8.
4. Mantel N, Haenszel W. Statistical aspects of the analysis of data from retrospective studies of disease. *J Natl Cancer Inst* 1959;**22**(4):719–48.
5. Rothman KJ. *Epidemiology: an introduction.* 2nd ed. New York: Oxford University Press; 2012.
6. Karlberg J, Chong DSY, Lai WYY. Do men have a higher case fatality rate of severe acute respiratory syndrome that women do? *Am J Epidemiol* 2004;**159**(3):229–31.
7. Stark CR, Mantel N. Effects of maternal age and birth order on the risk of mongolism and leukemia. *J Natl Cancer Inst* 1966;**37**(5):687–98.
8. Doll R, Hill AB. Mortality of British doctors in relation to smoking: observations on coronary thrombosis. In: Haenszel W, editor. *Epidemiological approaches to the study of cancer and other chronic diseases. National Cancer Institute monograph 19.* Bethesda: U.S. Department of Health, Education, and Welfare; 1966.

CHAPTER

Confounding in Cohort Studies - Part I

8

Confounding is a methodologic issue of fundamental importance in epidemiology. Although not pointed out at the time, we have already encountered an example of confounding. In our analysis of all-causes mortality in Alaska and Florida in Chapter 5, we found that differing age distributions in the two states affected the mortality comparison. More specifically, our initial analysis based on crude mortality rates appeared to show higher mortality in Florida, but after accounting for differences in the age distributions of Alaska and Florida we found greater overall mortality in Alaska. As will follow from the discussion below, "age" is a confounder of the association between "place of residence" and "all-causes mortality".

It may come as a surprise to learn that there is more than one definition of "confounding". In this chapter, three definitions are considered. The *classical* definition is intuitively appealing and has been a feature of epidemiology for decades. The more recent *causal diagram* and *counterfactual* definitions are much more sophisticated, but come at the cost of greater abstraction and statistical complexity. We will see that in simple cases the three definitions are closely related.

As we proceed, we will seek to identify "overall" measures of effect, that is, measures of effect that are unconfounded (according to a given definition of confounding) and which summarize an exposure-disease association over the values of a stratifying variable. Under various conditions, one or more of the crude, standardized and Mantel-Haenszel measures of effect can function in this role. It must be emphasized that the existence of overall measures of effect does not exclude stratum-specific measures of effect from further consideration. On the contrary, overall and stratum-specific measures of effect complement each other when the findings from an epidemiologic study are being reported.

8.1 Causal Diagram Confounding

Causal diagrams are used to encode in a pictorial manner the associations that exist among a collection of variables.[1(p.183–209), 2, 3(p.208–216)] Fig. 8.1 is a causal diagram for coronary heart disease. Hypertension, dyslipidemia, diabetes, smoking, heredity, age and gender are all direct causes of atherosclerosis, in the sense that they do not operate through intermediaries (at least not through ones that are shown in the causal diagram). Inactivity, central obesity and unhealthy diet are indirect causes of atherosclerosis.

Epidemiologic Methods. https://doi.org/10.1016/B978-0-44-318780-3.00014-2
Copyright © 2023 Elsevier Inc. All rights reserved.

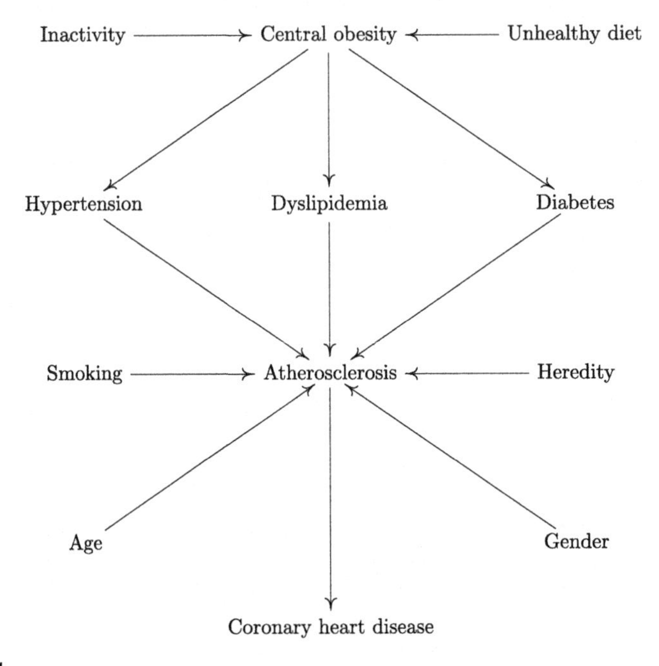

FIGURE 8.1

Causal diagram for coronary heart disease.

More generally, a *causal diagram* is a collection of variables connected by arrows. In what follows, $\mathcal{C}$ denotes an arbitrary causal diagram, and X, Y and Z are arbitrary variables in $\mathcal{C}$. Much terminology needs to defined before we can speak with precision about causal diagrams.

A *path* between X and Y is any nonrepeating sequence of arrows and variables that has X at one end and Y at the other, regardless of the direction of the arrows. In Fig. 8.5, $G \rightarrow F \rightarrow D$ and $G \rightarrow F \leftarrow H \rightarrow D$ are paths between G and D. A path between X and Y in which all the arrows are pointing away from X (and toward Y) is called a *causal path* (or *directed path*) from X to Y; that is, the path is of the form $X \rightarrow \cdots \rightarrow \cdots \rightarrow Y$. A path that is not a causal path is referred to as a *noncausal path* (or *undirected path*). In Fig. 8.5, $G \rightarrow E \rightarrow D$ and $G \rightarrow F \rightarrow E \rightarrow D$ are causal paths from G to D, but $G \rightarrow F \leftarrow H \rightarrow D$ is not. The definition of a causal path has a clear sense of directionality, which accounts for the change in terminology from "between X and Y" to "from X to Y". If there is a causal path from X to Y, we say that X is a *cause* (or *ancestor*) of Y, and that Y is an *effect* (or *descendant*) of X. When a causal path from X to Y consists of a single arrow, that is, the path is $X \rightarrow Y$, we say that X is a *direct cause* (or *parent*) of Y, and that Y is a *direct effect* (or *child*) of X. We say that X and Y are *neighbors* (or *adjacent*) if X is a direct cause of Y, or Y is a direct cause of X. In Fig. 8.5, G and F are neighbors, but G and D are not.

When constructing a causal diagram, it is not necessary to include all the causes of all the variables in the causal diagram. But if two or more variables are known to have a common cause, the causative variable must be shown in the causal diagram. We say that a causal diagram is *acyclic* if there are no causal paths that begin and end at the same variable; that is, no variable is a cause of itself. We henceforth assume that all causal diagrams are acyclic.

A variable on a causal path but not at either end of the path is said to be a *mediator* (or *intermediate*) on the path. See Fig. 8.2. A variable X on a path is said to be a *fork* on the path if it has two neighbors on the path that are direct effects of X; that is, the path has the form $\cdots Y \leftarrow X \rightarrow Z \cdots$. See Fig. 8.3. We say that a variable X on a path is a *collider* on the path, and that the path *collides* at X, if X has two neighbors on the path that are direct causes of X; that is, the path has the form $\cdots Y \rightarrow X \leftarrow Z \cdots$. See Fig. 8.4. If X is not a collider on a path, it is said to be a *noncollider* on the path. We say that a path is *open* (or *unblocked*) if it has no colliders, otherwise it is said to be *closed* (or *blocked*). In Fig. 8.5, the path $G \rightarrow F \leftarrow H$ collides at F, but the paths $G \rightarrow F \rightarrow D$ and $E \leftarrow F \rightarrow D$ do not; so F is a collider on the first path, but not on the other two paths.

A path between X and Y is called a *backdoor path* from X to Y if it is noncausal and begins with an arrow pointing into X. The definition has a sense of directionality, which justifies the terminology "from X to Y". In Fig. 8.5, $E \leftarrow F \rightarrow D$ is a backdoor path from E to D and also a backdoor path from D to E.

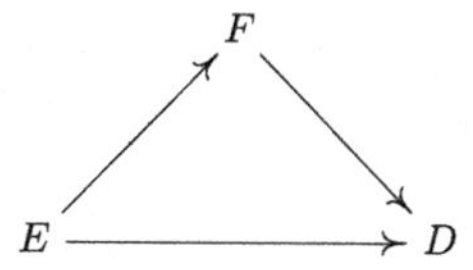

FIGURE 8.2

Causal diagram with F a mediator.

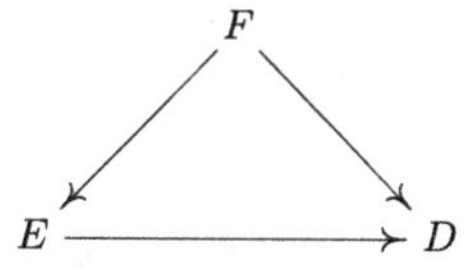

FIGURE 8.3

Causal diagram with F a fork.

Suppose that a closed cohort study of an exposure E and disease D has been conducted with the aim of estimating an overall measure of effect for the E-D association. Suppose further that the causal diagram $\mathcal{C}$ accurately depicts the associations among study variables. It can be shown that for there to be a crude association between E and D in the study data, there must be one or more open paths between E and D in the causal diagram.[2,4]

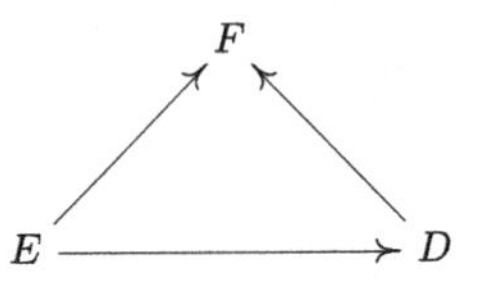

FIGURE 8.4

Causal diagram with F a collider.

Let $\mathcal{S}$ be a nonempty set of categorical variables in $\mathcal{C}$ other than E and D. The strata formed by cross-classifying the variables in $\mathcal{S}$ will be called the *strata of* $\mathcal{S}$. For example, if there are two variables, say age group and gender, the strata are given by all age group × gender categories. As will be discussed in subsequent sections, one of the ways to control confounding by the variables in $\mathcal{S}$ involves stratifying on them in the data analysis. There is a corresponding process in causal diagrams that we refer to as *conditioning* on $\mathcal{S}$. The key feature of conditioning on $\mathcal{S}$ is that it has the potential to change the pattern of associations in $\mathcal{C}$. In particular, conditioning on $\mathcal{S}$ might change an open path in $\mathcal{C}$ to a "closed" path, or change a closed path in $\mathcal{C}$ to an "open" path. The terms "open" and "closed" are written in quotations to indicate that they refer to paths in $\mathcal{C}$ after conditioning has been implemented.

The easiest way to grasp what "open" and "closed" mean is to consider a few illustrative examples. Let us investigate the effect of conditioning on $\{F\}$, the set consisting only of F, in Figs. 8.2–8.4. In Fig. 8.2, $E \rightarrow F \rightarrow D$ is a causal path from E to D. Conditioning on $\{F\}$ "closes" the (open) path. Stratification by F in the data analysis is undesirable because it disrupts the causal association between E and D through the mediator F.

In Fig. 8.3, $E \leftarrow F \rightarrow D$ is an open backdoor path from E to D. Since F is a common cause of E and D, it follows that E and D are associated. For example, cigarette smoking is a cause of both lung cancer and coronary heart disease. Someone diagnosed with lung cancer is more likely than average to be a smoker, and therefore more likely than average to have coronary heart disease. This shows that lung cancer and coronary heart disease are associated, even though neither disease is a cause of the other. Thus, the path $E \leftarrow F \rightarrow D$ represents a noncausal association between E and D. Conditioning on $\{F\}$ "closes" the (open) path. Stratification by F in the data analysis is desirable because it disrupts the noncausal association between E and D through the fork F.

In Fig. 8.4, $E \rightarrow F \leftarrow D$ is a closed path between E and D. Even though F is a common effect of E and D, it is nevertheless true that E and D are not associated. For example, diverse diseases cause people to seek medical attention, and yet most diseases are unrelated to each other. However, as unintuitive as it may seem, conditioning on $\{F\}$ "opens" the (closed) path. To see this, let us examine stratification by F in the extreme situation where E and D are the only causes of F, and all three variables are dichotomous. Consider the stratum where F is present. If E is absent, then D must be present; and if D is absent, then E must be present. Thus, E and D are negatively associated in this stratum. Stratification by F in the data analysis is undesirable because it creates a noncausal association between E and D through the collider F, and can lead to what is termed *collider bias*.[5]

Returning to the general setting and expressing ourselves loosely, we can say that "confounding" of the E-D association is present if causal associations between E and D are obscured by noncausal associations between E and D. As illustrated by Fig. 8.3, noncausal associations in a causal diagram are due to open backdoor paths. This provides a rationale for the following definition. We say there is *causal diagram confounding* of the E-D association if and only if the following criterion is satisfied[1(p.193), 2, 3(p.208–216), 6(p.84), 7]:

(C1) There is an open backdoor path from E to D.

(The definition of causal diagram confounding varies somewhat in the epidemiologic literature.)

In Fig. 8.3, $E \leftarrow F \rightarrow D$ is an open backdoor path from E to D, and so there is causal diagram confounding of the E-D association. In Fig. 8.2, there is no open backdoor path from E to D, hence no causal diagram confounding of the E-D association; and likewise for Fig. 8.4.

We say that $\mathcal{S}$ is *sufficient* to control causal diagram confounding of the E-D association if there is no causal diagram confounding of the E-D association in any of the strata of $\mathcal{S}$.[1(p.192), 2, 3(p.208–216)] A sufficient set is said to be *minimally sufficient* if no proper subset is sufficient. A minimally sufficient set can be obtained from a sufficient set by sequentially dropping variables until the set is no longer sufficient. Depending on which variables are dropped, different minimally sufficient sets may result. We say that a variable is a *causal diagram confounder* if it appears in at least one minimally sufficient set. When there is no causal diagram confounding, it is convenient to say that the empty set, denoted here by $\{\}$, is sufficient.

It can be shown that $\mathcal{S}$ is sufficient to control causal diagram confounding of the E-D association if and only if the following conditions, known collectively as the *backdoor criterion*, are satisfied[1(p.192), 2, 3(p.208–216), 4, 6(p.85), 7, 8(p.61–64)]:

(C2) E is not a cause of any variable in $\mathcal{S}$.

(C3) There is no "open" backdoor path from E to D after conditioning on $\mathcal{S}$.

It was pointed out above that there is causal diagram confounding of the E-D association in Fig. 8.3, but not in Figs. 8.2 and 8.4. So $\{\}$ is a (minimally) sufficient set for Figs. 8.2 and 8.4. For Fig. 8.3, let $\mathcal{S} = \{F\}$. Since E is not a cause of F and the only open backdoor path from E to D contains F, (C2) and (C3) are satisfied. So $\{F\}$ is a (minimally) sufficient set.

Let us now examine two causal diagrams that are more complicated. In Fig. 8.5, $E \leftarrow F \rightarrow D$, $E \leftarrow G \rightarrow F \rightarrow D$ and $E \leftarrow F \leftarrow H \rightarrow D$ are open backdoor paths from E to D, and so causal diagram confounding of the E-D association is present. There is also the backdoor path $E \leftarrow G \rightarrow F \leftarrow H \rightarrow D$ from E to D, but it is closed (because F is a collider on the path). Let $\mathcal{S} = \{F\}$. Since E does not cause F, (C2) is satisfied. If we condition on $\{F\}$, the open paths $E \leftarrow F \rightarrow D$, $E \leftarrow G \rightarrow F \rightarrow D$ and $E \leftarrow F \leftarrow H \rightarrow D$ become "closed" (because F is a noncollider on each of them), but the closed backdoor path $E \leftarrow G \rightarrow F \leftarrow H \rightarrow D$ becomes "open" (because F is a collider on the path). So (C3) is not satisfied, hence $\{F\}$ is not a sufficient set. Now let $S = \{F, G\}$. Since E does not cause F or G, (C2) is satisfied. If we condition on $\{F, G\}$, the open paths $E \leftarrow F \rightarrow D$, $E \leftarrow G \rightarrow F \rightarrow D$ and $E \leftarrow F \leftarrow H \rightarrow D$ become "closed" as before, but this time the closed path $E \leftarrow G \rightarrow F \leftarrow H \rightarrow D$ becomes "closed" (because, although F is a collider on the path, G is a noncollider on the path). So (C3) is satisfied, hence $\{F, G\}$ is a (minimally) sufficient set. Similarly, $\{F, H\}$ is a (minimally) sufficient set and $\{F, G, H\}$ is a sufficient set.

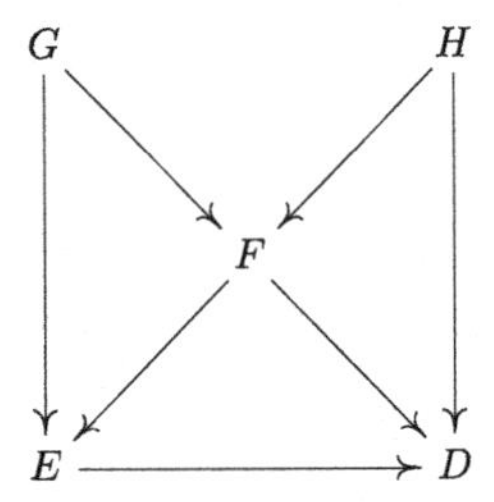

FIGURE 8.5

Bow tie-shaped causal diagram.

In Fig. 8.6, there are no open backdoor paths from E to D, and so causal diagram confounding of the E-D association is absent. There is the backdoor path $E \leftarrow G \rightarrow F \leftarrow H \rightarrow D$ from E to D, but it is closed (because F is a collider on the path). So $\{\}$ is a (minimally) sufficient set. If we condition in turn on $\{F\}$, $\{F, G\}$, $\{F, H\}$ and $\{F, G, H\}$, the same arguments used in connection with Fig. 8.5 show that $\{F\}$ is not a sufficient set, but $\{F, G\}$, $\{F, H\}$ and $\{F, G, H\}$ are. This demonstrates that conditioning on $\{F\}$ introduces confounding where there was none.

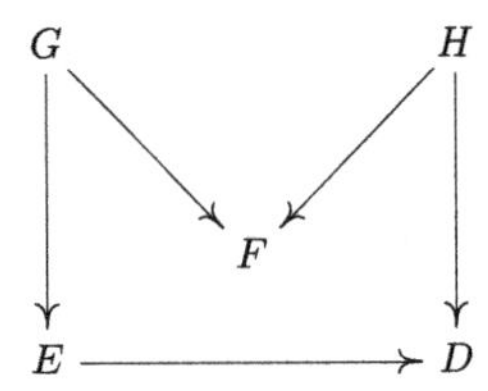

FIGURE 8.6

M-shaped causal diagram.

It was pointed out in connection with Fig. 8.4 that conditioning on a collider can be a source of bias. However, the existence of colliders does not always spell trouble. In the following example we return to a discussion of the birth weight paradox discussed in Section 6.5 and demonstrate that an understanding of colliders can be helpful in explaining a perplexing phenomenon.

Example: Birth Weight Paradox

We saw in Section 6.5 that birth weight is a qualitative effect modifier of the association between smoking status of the mother during pregnancy and infant mortality. It is known that maternal smoking during pregnancy is associated with low birth weight and increased infant mortality, and likewise that serious congenital anomalies are associated with low birth weight and increased infant mortality.[9,10] Treating these associations as causal produces Fig. 8.7.

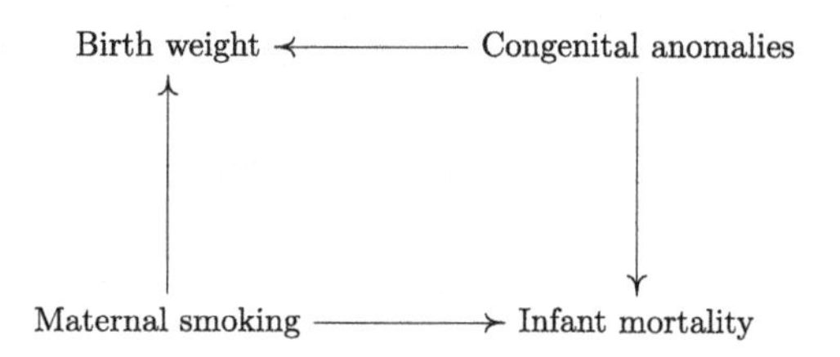

FIGURE 8.7

Causal diagram for birth weight paradox.

Observe that Birth weight is a collider on the path Maternal smoking → Birth weight ← Congenital anomalies. The argument used above in connection with Fig. 8.4 shows that in low birth weight newborns, maternal smoking during pregnancy is associated with a decrease in the congenital anomaly rate. (Note that it is not being asserted that maternal smoking causes a reduction in the congenital anomaly rate.) If Fig. 8.7 is a valid depiction of the associations among the variables under consideration, it follows that in low birth weight newborns, maternal smoking during pregnancy is associated with a decrease in infant mortality. Thus, stratification on a collider may provide an explanation for the birth weight paradox.[10,11] ◊

8.2 Counterfactual Confounding in Closed Cohort Studies

For the remainder of this chapter (except the final section), we restrict ourselves to the closed cohort study described in Section 6.1, where only three variables are considered—an exposure variable E, a disease variable D, and a stratifying variable F. Both E and D are dichotomous with values 0,1, and F is polychotomous with values $0, \ldots, k$. We sometimes refer to F as a *potential confounder*. Whether F turns out to be an actual confounder depends, in part, on the definition of confounding adopted.

The counterfactual definition of confounding is based on the concept of *potential outcomes* in a *target population*, where possibly different outcomes can occur to members of the target population depending on whether they are exposed or not. (Our use of the term "target population" in the present context differs from its use in Section 2.6.) For the present discussion, we take the target population to be the members of the exposed cohort. In that case, potential outcomes following exposure are realized and the risk of becoming a case is R_1, as observed. Under the hypothetical scenario that members of the exposed cohort are not exposed, the corresponding risk is denoted by $R_1^{\bullet}$. We say that $R_1^{\bullet}$ is a *counterfactual risk* because it is based on an imaginary situation that is contrary to fact.

The exposed cohort in the absence of exposure is the ideal comparison group for the exposed cohort (in the presence of exposure). Any difference between R_1 and $R_1^{\bullet}$ must be due to E. We are led to consider the corresponding measures of effect. The *(crude) counterfactual risk ratio*, *(crude) counterfactual odds ratio* and *(crude) counterfactual risk difference* are defined in a manner analogous to their counterparts in (4.3), except that $R_1^{\bullet}$ is used in place of R_0 [12]:

$$RR^{\bullet} = \frac{R_1}{R_1^{\bullet}}$$

$$OR^{\bullet} = \frac{R_1/(1 - R_1)}{R_1^{\bullet}/(1 - R_1^{\bullet})} \tag{8.1}$$

$$RD^{\bullet} = R_1 - R_1^{\bullet}.$$

By definition, $RR^{\bullet}$, $OR^{\bullet}$ and $RD^{\bullet}$ have a causal interpretation.

We say there is *counterfactual confounding* of the E-D association if and only if the following condition is satisfied [12–17]:

$$\text{(C4)} \quad R_0 \neq R_1^{\bullet}.$$

That is, counterfactual confounding is present precisely when the risk of becoming a case in the unexposed cohort is different from the risk of becoming a case in the exposed cohort in the absence of exposure. It is intuitively clear that for the closed cohort study we are considering, a difference in the values of R_0 and $R_1^{\bullet}$ must be due to F. When (C4) is true, we say that F is a *counterfactual confounder* of the E-D association.

However, working with (C4) in practice presents an obvious problem: $R_1^{\bullet}$ is the risk of becoming a case in an imaginary cohort. In order to perform an actual data analysis, we make the following assumptions:

(C5) E is not a cause of F.

(C6) There is no counterfactual confounding of the E-D association in the strata of F.

As we will soon see, using (C5) and (C6) we are able to determine whether (C4) is true using only study data. However, deciding on the validity of (C5) and (C6) depends on having prior knowledge of the causal associations among E, D and F, and possibly other variables. If E is a well-studied exposure, it might be possible to make a decision about (C5). But (C6) is another story. Since we are trying to determine if there is counterfactual confounding of the E-D association in the overall cohort, it seems unlikely that adequate background information would be available on whether there is counterfactual confounding of the E-D association in the strata of F. In particular, there may be variables, known as *unmeasured confounders*, that produce such confounding and yet are unknown to the investigators. Thus, (C6) is a crucial assumption that can never be fully justified.

There is an obvious similarity between (C5)-(C6) and the backdoor criterion with $\mathcal{S} = \{F\}$. If we have the prior knowledge to determine the validity of (C5) and (C6), we are presumably in a position to construct the causal diagram corresponding to the closed cohort study. In that case, why bother with (C4), or for that matter, counterfactual methods in general? The answer is that counterfactual techniques go beyond the qualitative approach inherent in causal diagrams to encompass quantitative methods. In particular, they provide a way to compute overall measures of effect.

It is shown in Appendix G that assumptions (C5) and (C6) imply[4, 6(p.19), 8(p.55–59), 12, 14, 17, 18]

$$R_1^\bullet = R_{0(1)}\,, \tag{8.2}$$

from which it follows that (C4) is equivalent to

$$\text{(C7)} \quad R_0 \neq R_{0(1)}\,,$$

where $R_{0(1)}$, the standardized risk in the unexposed cohort, is defined in (7.3). Thus, F is a counterfactual confounder if and only if $R_0 \neq R_{0(1)}$, and this can be determined using study data. It can be shown that

$$\begin{aligned} R_0 \neq R_{0(1)} \quad & \text{if and only if} \quad RR \neq sRR \\ & \text{if and only if} \quad OR \neq sOR \\ & \text{if and only if} \quad RD \neq sRD\,, \end{aligned}$$

which provides three alternatives to (C7):

$$\text{(C8)} \quad RR \neq sRR \quad \text{or} \quad OR \neq sOR \quad \text{or} \quad RD \neq sRD\,.$$

It follows from (7.4), (8.1) and (8.2) that the crude counterfactual measures of effect equal their standardized counterparts:

$$\begin{aligned} RR^\bullet &= sRR \\ OR^\bullet &= sOR \\ RD^\bullet &= sRD\,. \end{aligned}$$

Since the crude counterfactual measures of effect are counterfactual unconfounded (by definition), the standardized measures of effect are also counterfactual unconfounded. This means that the standardized measures of effect are suitable as overall measures of effect.

Let us denote by $a_1^\bullet$ the number of incident cases in the exposed cohort in the absence of exposure. It is shown in Appendix G that

$$a_1^\bullet = a_{0(1)}\,, \tag{8.3}$$

where $a_{0(1)}$, the expected number of incident cases in the exposed cohort, is defined in (7.5).

8.3 Classical Confounding in Closed Cohort Studies

The causal diagram definition of confounding is concerned with the existence of paths from exposure to disease that give the appearance of causation when none exists. The classical definition of confounding is aimed at the same issue but framed in more elementary terms. From this perspective, the classical definition can be viewed as a forerunner of the causal diagram definition. For this reason, it is often helpful to employ causal diagrams, albeit simple ones, in discussions of classical confounding.

We say that F is a *classical confounder* of the E-D association if and only if the following conditions are satisfied[1(p.129–134), 19]:

(C9) Neither E nor D is a cause of F.

(C10) F is associated with E in the cohort at the start of follow-up.

(C11) F is a risk factor for D in the unexposed cohort.

When a variable is a classical confounder, we say that *classical confounding* is present. An alternative to (C11) that appears in the literature is:

(C11′) F is a cause of D in the unexposed cohort.

Since a risk factor can be a cause of the disease or merely associated with the disease, (C11′) is a more restrictive criterion than (C11). Unless stated otherwise, we will stay with (C11) throughout.

Let us temporarily replace (C11) with (C11′) and see how the classical and causal diagram definitions of confounding compare in Figs. 8.2–8.4. The condition "at the start of follow-up" in (C10) cannot be encoded in a causal diagram, and so we ignore this aspect of (C10) in the following discussion. In Fig. 8.2, E is a cause of F; and in Fig. 8.4, E and D are causes of F. In Fig. 8.3, neither E nor D cause F, F is associated with E (because it is a cause of E), and F is a risk factor for D in the unexposed cohort (because it is a direct cause of D). This shows that F is a classical confounder of the E-D association in Fig. 8.3, but not in Figs. 8.2 and 8.4. We saw in Section 8.1 that there is causal diagram confounding of the E-D association in Fig. 8.3, but not in Figs. 8.2 and 8.4. Thus, the classical and causal diagram definitions of confounding agree in Figs. 8.2–8.4.

We now return to the definition of classical confounding that uses (C11) and consider how the classical and causal diagram definitions of confounding compare in

Fig. 8.6. As in the previous comparisons, the condition "at the start of follow-up" in (C10) is ignored. In Fig. 8.6, neither E nor D cause F, F is associated with E (because F and E have the common cause G), and F is associated with D in the unexposed cohort (because F and D have the common direct cause H). This shows that F is a classical confounder of the E-D association. However, we saw in Section 8.1 that there is no causal diagram confounding of the E-D association in Fig. 8.6. Thus, we have a conflict between the causal diagram and classical definitions of confounding.

Conditions (C9)–(C11) can be evaluated on the basis of prior knowledge of the associations between E, D and F. We now show that (C10) and (C11) can be recast in mathematically explicit terms, thereby allowing them to be assessed using study data rather than having to rely on background knowledge. This approach to evaluating (C10) and (C11) would rarely be performed in practice, but it is instructive in the present context. Expressed as it is in causal terms, (C9) is beyond the reach of standard methods of data analysis. We first observe that F is associated with E in the cohort at the start of follow-up precisely when the distribution of F differs in the exposed and unexposed cohorts at the start of follow-up, that is, when $n_{1i}/n_1 \neq n_{0i}/n_0$ for some value (but not necessarily all values) of i. We next observe that F is a risk factor for D in the unexposed cohort precisely when the risk of becoming a case varies over the strata of F, that is, when $R_{00}, \ldots, R_{0i}, \ldots, R_{0k}$ are not all equal. It follows that (C9)–(C11) can be replaced with:

(C12) Neither E nor D is a cause of F.

(C13) $\dfrac{n_{1i}}{n_1} \neq \dfrac{n_{0i}}{n_0}$ for some value of i.

(C14) $R_{00}, \ldots, R_{0i}, \ldots, R_{0k}$ are not all equal.

If F is not a classical confounder of the E-D association, the crude measures of effect are unconfounded (by definition), making them suitable as overall measures of effect. On the other hand, if F is a classical confounder, we have a problem: the classical approach to confounding does not automatically provide choices for overall measures of effect. Historically, the Mantel-Haenszel measures of effect have been used for this purpose, even when the homogeneity assumption is difficult to justify.

8.4 Counterfactual and Classical Confounding in Closed Cohort Studies

Example: Confounding in Closed Cohort Studies #1–3

We now illustrate the above theory using closed cohort studies #1–3 presented in Section 7.1. Tables 7.6(c), 7.7(c) and 7.8(c) provide the information needed to assess counterfactual confounding using (C8). Tables 7.6(a), 7.7(a) and 7.8(a) and Tables 7.6(b), 7.7(b) and 7.8(b) provide the information needed to evaluate classical confounding using (C13) and (C14). Table 8.1 summarizes the findings for the two definitions of confounding. ◊

Table 8.1 Confounding in closed cohort studies #1–3 (present: +; absent: −).

Study	Counterfactual	Classical
#1	−	−
#2	+	+
#3	−	+

Example: Confounding in SARS and Down Syndrome Studies

We now consider confounding in the SARS and Down syndrome studies, using the same approach as in the preceding example. See Tables 7.9(a)–(c) and 7.10(a)–(c). It is evident that there is counterfactual confounding and classical confounding in the Down syndrome study, but the situation for the SARS study is less clear on both counts.

Let us illustrate the epidemiologic interpretation of counterfactual confounding in the Down syndrome study, taking the risk ratio to be the measure of effect. The aim of the study was to examine whether birth order is a risk factor for Down syndrome. The crude risk ratio of 1.86 indicates that an association exists. Older maternal age was known at the time of the study to be a risk factor for Down syndrome and, not surprisingly, maternal age correlates with birth order. The standardized risk ratio of 1.04 shows that after controlling for maternal age, the association between birth order and Down syndrome disappears. We conclude that maternal age is a counterfactual confounder of the birth order-Down syndrome association, and that after controlling for maternal age, birth order is not a risk factor for Down syndrome. ◊

Although the counterfactual and classical definitions of confounding are outwardly very different, they have a surprising degree of overlap, as we now show.[20] For the counterfactual definition, we implicitly assume that (C5) is satisfied. Let us level the playing field by assuming that for the classical definition, (C9) is satisfied. This means that the counterfactual definition is determined by (C7), and the classical definition is determined by (C10) and (C11), or equivalently by (C13) and (C14).

With these assumptions, it is shown in Appendix G that:

1. If F is a *counterfactual* confounder, then it is a *classical* confounder.
2. If F is dichotomous, then it is a *counterfactual* confounder if and only if it is a *classical* confounder.

In contrast to item 2, we find from closed cohort study #3 in Table 8.1 that:

3. If F is not dichotomous, then it can be a *classical* confounder but not a *counterfactual* confounder.

It is readily verified that Table 8.1 is consistent with items 1–3. Item 1 implies that (C10) and (C11) are necessary conditions for F to be a counterfactual confounder. Item 2 implies that when F is dichotomous, (C10) and (C11) are necessary and sufficient for F to be a counterfactual confounder.

8.5 Collapsibility in Closed Cohort Studies

We say that a measure of effect is *collapsible* over F if its crude value equals a weighted average of stratum-specific values, for some set of weights; otherwise the measure of effect is said to be *noncollapsible* over F.[6(p.54), 15, 21(p.193)] Let us denote the crude value of an arbitrary measure of effect by M, and the minimum and maximum stratum-specific values by M_{min} and M_{max}, respectively.

It is shown in Appendix G that:

1. A measure of effect is collapsible over F if and only if its crude value falls within the range spanned by the stratum-specific values, that is, if and only if $M_{min} \leq M \leq M_{max}$.
2. If F is not a counterfactual confounder, then the risk ratio and risk difference are collapsible over F.

Based on item 1, Table 8.2 gives the collapsibility/noncollapsibility status of the risk ratio, odds ratio and risk difference in closed cohort studies #1–3. In closed cohort study #3, all the measures of effect are collapsible over F, while in closed cohort study #2, none of them are. Collapsibility is not necessarily an all-or-nothing phenomenon. In closed cohort study #1, the risk ratio and risk difference are collapsible over F but the odds ratio is not.

Table 8.2 Collapsibility in closed cohort studies #1–3 (collapsible: +; noncollapsible: −).

Study	Risk ratio	Odds ratio	Risk difference
#1	+	−	+
#2	−	−	−
#3	+	+	+

We observe from Tables 8.1 and 8.2 that item 2 is satisfied for closed cohort studies #1 and 3. In contrast to item 2, we find from closed cohort study #1 in Tables 8.1 and 8.2 that:

3. If F is not a counterfactual confounder, the odds ratio may be noncollapsible over F.

The noncollapsibility of the odds ratio leads to difficulties in the interpretation of study #1 when we adopt the counterfactual approach to confounding and focus on the odds ratio as the measure of effect. To make matters concrete, suppose stratification is by gender. Then the odds ratio for males is $hOR = 6.0$, and likewise for females. But the overall odds ratio for males and females combined is $OR = 3.6$ because, according to Table 8.1, gender is not a counterfactual confounder. This highly unintuitive finding is due to the noncollapsibility of the odds ratio and raises concerns about the suitability of the odds ratio as a measure of effect. It follows from item 2 that a similar problem does not occur with the risk ratio or risk difference.

The odds ratio is not alone in having collapsibility problems. In open cohort studies, both the incidence rate ratio and incidence rate difference are noncollapsible.[22]

8.6 Choosing a Definition of Confounding

Based on the attention it receives in the current epidemiologic literature, there is little question that the counterfactual definition of confounding is considered to be "the" definition of confounding, at least among epidemiologic methodologists. Causal diagrams play an important role in the counterfactual approach to confounding. At a minimum, they can be used to encode what is known or suspected about causal connections among variables relevant to the exposure-disease association under consideration. If prior knowledge is extensive, a set of variables sufficient to control counterfactual confounding can be identified and incorporated into a data analysis. When background knowledge is scanty and constructing a detailed causal diagram is not possible, the classical approach to confounding can be called upon. As remarked in Section 8.4, conditions (C10) and (C11), which form part of the classical definition of confounding, must be satisfied for a variable to be a counterfactual confounder. We can use these conditions to identify potential counterfactual confounders when a study is being designed.[23]

The counterfactual definition of confounding is appealing due to its solid philosophical and mathematical foundations, as well as its ability to handle difficult problems in the statistical analysis of epidemiologic data, such as time-dependent confounding.[6(p.235–276)] However, counterfactual methods are often theoretically and computationally complex.[6] For that reason, we will not explore the counterfactual approach to confounding beyond what has already been presented for closed cohort studies.

8.7 Classical Confounding in Stationary Populations

Let us continue with the notation for stationary populations presented in Chapter 5, but instead of focusing exclusively on the stratifying variable age, we now let the polychotomous variable F be an arbitrary potential confounder. We say that F is a *classical confounder* of the E-D association if and only if the following conditions are satisfied:

(C15) Neither E nor D is a cause of F.

(C16) F is associated with E in the population.

(C17) F is a risk factor for D in the unexposed population.

In keeping with (C9)–(C11) and (C12)–(C14), we recast (C15)–(C17) in mathematically explicit terms as follows:

(C18) Neither E nor D is a cause of F.

(C19) $\frac{M_{1i}}{M_1} \neq \frac{M_{0i}}{M_0}$ for some value of i.

(C20) $I_{00}, \ldots, I_{0i}, \ldots, I_{0k}$ are not all equal.

Example: Alaska-Florida Mortality Comparison

Returning to the comparison of mortality in Alaska and Florida that introduced this chapter, let E, D and F represent "place of residence", "all-causes mortality" and "age", respectively. Evidently, where people live and their eventual death do not affect their current age, and so (C18) is satisfied. We have from Table 5.1(c) that (C19) is satisfied, and from Table 5.1(b) that so is (C20). This shows that age is a classical confounder of the association between place of residence and all-causes mortality. ◊

References

1. Rothman KJ, Greenland S, Lash TL, editors. *Modern epidemiology*. 3rd ed. Philadelphia: Lippincott Williams & Wilkins; 2008.
2. Greenland S, Pearl J, Robins JM. Causal diagrams for epidemiologic research. *Epidemiology* 1999;**10**(1):37–48.
3. Greenland S, Pearl J. Causal diagrams. In: Lovric M, editor. *International encyclopedia of statistical science, part 3*. Berlin: Springer-Verlag; 2011.
4. Pearl J. Causal diagrams for empirical research. *Biometrika* 1995;**82**(4):669–710.
5. Cole SR, Platt RW, Schisterman EF, et al. Illustrating bias due to conditioning on a collider. *Int J Epidemiol* 2010;**39**(2):417–20.
6. Hernán MA, Robins JM. *Causal inference: what if*. Boca Raton: Chapman & Hall/CRC; 2020. https://www.hsph.harvard.edu/miguel-hernan/causal-inference-book.
7. Hernán MA. Confounding. In: Melnick EL, Everitt BS, editors. *Encyclopedia of quantitative risk analysis and assessment*. Chichester: Wiley; 2008.
8. Pearl J, Glymour M, Jewell NP. *Causal inference in statistics: a primer*. Chichester: John Wiley & Sons; 2016.
9. Wilcox AJ. On the importance—and the unimportance—of birthweight. *Int J Epidemiol* 2001;**30**(6):1233–41.
10. Hernández-Díaz S, Wilcox AJ, Schisterman EF, et al. From causal diagrams to birth weight-specific curves of infant mortality. *Eur J Epidemiol* 2008;**23**(3):163–6.
11. Hernández-Díaz S, Schisterman EF, Hernán MA. The birth weight "paradox" uncovered? *Am J Epidemiol* 2006;**164**(11):1115–20.

12. Wickramaratne PJ, Holford TR. Confounding in epidemiologic studies: the adequacy of the control group as a measure of confounding. *Biometrics* 1987;**43**(4):751–65.
13. Greenland S, Morgenstern H. Confounding in health research. *Annu Rev Public Health* 2001;**22**:189–212.
14. Greenland S, Robins JM. Identifiability, exchangeability, and epidemiological confounding. *Int J Epidemiol* 1986;**15**(3):412–8.
15. Greenland S, Robins JM, Pearl J. Confounding and collapsibility in causal inference. *Stat Sci* 1999;**14**(1):29–46.
16. Maldonado G, Greenland S. Estimating causal effects. *Int J Epidemiol* 2002;**31**(2):422–9.
17. Robins JM, Morgenstern H. The foundations of confounding in epidemiology. *Comp Math Appl* 1987;**14**(9–12):869–916.
18. Hernán MA, Robins JM. Estimating causal effects from epidemiological data. *J Epidemiol Community Health* 2006;**60**(7):578–86.
19. Miettinen OS, Cook EF. Confounding: essence and detection. *Am J Epidemiol* 1981;**114**(4):593–603.
20. Newman SC. Commonalities in the classical, collapsibility and counterfactual concepts of confounding. *J Clin Epidemiol* 2004;**57**(4):325–9.
21. Pearl J. *Causality: models, reasoning, and inference*. 2nd ed. Cambridge: Cambridge University Press; 2009.
22. Greenland S. Absence of confounding does not correspond to collapsibility of the rate ratio or rate difference. *Epidemiology* 1996;**7**(5):498–501.
23. VanderWeele TJ. Principles of confounder selection. *Eur J Epidemiol* 2019;**34**(3):211–9.

CHAPTER

9 Confounding in Cohort Studies - Part II

9.1 Attributable Fractions in Closed Cohort Studies

We continue with the closed cohort study outlined at the beginning of Section 8.2. There are a_1 incident cases in the exposed cohort and, according to (7.5) and (8.3), there would be $a_{0(1)}$ incident cases in the exposed cohort under the counterfactual condition of no exposure. Let us assume that exposure increases the risk of becoming a case, so that $a_1 > a_{0(1)}$. Then the number of incident cases in the exposed cohort that would be prevented if the exposure were eliminated, that is, the number of incident cases in the exposed cohort attributable to exposure, is $a_1 - a_{0(1)}$.

The *(standardized) exposed attributable fraction* is defined to be the proportion of incident cases in the exposed cohort that would be prevented if the exposure were eliminated[1]:

$$sAF_E = \frac{a_1 - a_{0(1)}}{a_1}. \tag{9.1}$$

A related quantity is the *(standardized) population attributable fraction*, which is defined to be the proportion of incident cases in the entire cohort that would be prevented if the exposure were eliminated[1,2]:

$$sAF_P = \frac{a_1 - a_{0(1)}}{a}. \tag{9.2}$$

The definitions of sAF_E and sAF_P depend crucially on the causal nature of counterfactual methods.

It is shown in Appendix H that if there is no counterfactual confounding, then $a_{0(1)} = R_0 n_1$. In that case, (9.1) and (9.2) simplify to

$$\begin{aligned} AF_E &= \frac{a_1 - R_0 n_1}{a_1} \\ AF_P &= \frac{a_1 - R_0 n_1}{a}, \end{aligned} \tag{9.3}$$

which are referred to as the *(crude) exposed attributable fraction* and *(crude) population attributable fraction*, respectively.

The idea of attributable fraction was introduced decades ago.[3] Since then there has been a proliferation of terminology used to describe a variety of related indices,

Epidemiologic Methods. https://doi.org/10.1016/B978-0-44-318780-3.00015-4
Copyright © 2023 Elsevier Inc. All rights reserved.

and this can be a source of confusion. For example, the population attributable fraction has at various times been called the attributable risk, population attributable risk and etiologic fraction.[4]

Denote by f the proportion of all incident cases who are in the exposed cohort, by g the proportion of all subjects who are in the exposed cohort, and by h_i the proportion of all incident cases who are in stratum i:

$$f = \frac{a_1}{a}$$

$$g = \frac{n_1}{n} \tag{9.4}$$

$$h_i = \frac{a_{1i} + a_{0i}}{a}.$$

It is shown in Appendix H that[1–3,5,6]

$$AF_E = \frac{RR - 1}{RR} \tag{9.5}$$

$$AF_P = f \times AF_E = \frac{g(RR - 1)}{g(RR - 1) + 1} \tag{9.6}$$

$$sAF_E = \frac{sRR - 1}{sRR} \tag{9.7}$$

$$sAF_P = f \times sAF_E = \sum_{i=0}^{k} h_i \times AF_{Pi}, \tag{9.8}$$

where AF_{Pi} is the population attributable fraction for those members of the cohort in stratum i. Since $\sum_{i=0}^{k} h_i = 1$, we see that sAF_P is a weighted average of stratum-specific values.

Example: British Doctors' Study

In the 50-year follow-up of male physicians in the British Doctors' Study discussed in Section 4.2, the directly standardized mortality ratio for lung cancer comparing current cigarette smokers to lifelong nonsmokers is estimated to be $dMR = 14.7$.[7] Using this as an estimate of sRR, we have from (9.7) that

$$sAF_E = \frac{14.7 - 1}{14.7} = 89.1\%.$$

The public heath interpretation is that if a group of young men who are about to become lifelong cigarette smokers could be convinced not to take up the habit, about nine out of ten cases of lung cancer would be prevented. ◇

9.2 Unmeasured Confounders in Closed Cohort Studies

We continue with the closed cohort study of Section 9.1 and assume that F is a counterfactual confounder that is dichotomous. Suppose the risk ratio for the E-D association is homogeneous over F. Let us denote the crude risk ratio by cRR_E and the common stratum-specific value of the risk ratio by hRR_E. By assumption, cRR_E is counterfactual confounded. As discussed in Section 8.2, the standardized risk ratio is suitable as an overall measure of effect, and we have from (7.8) that its value equals hRR_E. It follows from (6.6) that the risk ratio for the F-D association is homogeneous over E. Let us denote the common stratum-specific value of the risk ratio by hRR_F.

Suppose data are not collected on F as part of the cohort study, perhaps because the investigators are unaware that F is a risk factor for D. Then cRR_E will be reported as the overall measure of effect instead of hRR_E. In this situation, F is said to be an *unmeasured confounder*. We take cRR_E/hRR_E to be a measure of the resulting bias.

Of the subjects in the cohort exposed to E at the start of follow-up, we denote by $p_{F=1|E=1}$ the proportion exposed to F at the start of follow-up. Similarly, of the subjects in the cohort not exposed to E at the start of follow-up, we denote by $p_{F=1|E=0}$ the proportion exposed to F at the start of follow-up. It is shown in Appendix H that[8–10]

$$\frac{cRR_E}{hRR_E} = \frac{p_{F=1|E=1}(hRR_F - 1) + 1}{p_{F=1|E=0}(hRR_F - 1) + 1}. \tag{9.9}$$

Table 9.1 gives values of cRR_E/hRR_E for selected values of hRR_F, $p_{F=1|E=1}$ and $p_{F=1|E=0}$. For given values of $p_{F=1|E=1}$ and $p_{F=1|E=0}$, the bias increases as

Table 9.1 cRR_E/hRR_E for selected values of hRR_F and $p_{F=1|E=1}$ and $p_{F=1|E=0}$.

$p_{F=1\|E=1}$	hRR_F			
	2	5	10	20
	$p_{F=1\|E=0} = 0.10$			
0.05	0.95	0.86	0.76	0.67
0.10	1.00	1.00	1.00	1.00
0.25	1.14	1.43	1.71	1.98
0.50	1.36	2.14	2.89	3.62
	$p_{F=1\|E=0} = 0.25$			
0.05	0.84	0.60	0.45	0.34
0.10	0.88	0.70	0.58	0.50
0.25	1.00	1.00	1.00	1.00
0.50	1.20	1.50	1.69	1.83

hRR_F increases, that is, as F becomes a stronger risk factor. We see that unmeasured confounding can result in substantial over- or underestimation of the standardized risk ratio.

9.3 Controlling Confounding in Cohort Studies

Methods to control confounding in a cohort study can be incorporated into the study design or implemented as part of the data analysis. At the design stage, the methods are *randomization*, *restriction* and *matching*. When the data are being analyzed, the options include *standardization*, *stratification* (e.g. Mantel-Haenszel methods), *regression* (e.g. Cox regression) and *inverse probability weighting*. Randomization [11] is considered in Chapter 11. Standardization [12] was discussed in Chapters 5, 7 and 8. Mantel-Haenszel methods [12] were introduced in Chapters 7 and 8. Cox regression [12,13] and inverse probability weighting [13–15] are beyond the scope of this book. That leaves restriction and matching, which are discussed next. We note in passing that standardization and inverse probability weighting are both conceptually related to design weighting discussed in Section 2.5.

Restriction

Restriction prevents confounding by intentionally limiting eligibility for recruitment into a cohort so that the confounder exhibits little or no variability in the cohort at the start of follow-up. For example, if gender is considered to be a confounder, restricting the study to either males or females eliminates the possibility of confounding by gender. Restriction is easy to implement and is inexpensive. Drawbacks are that the option of evaluating the confounder as a risk factor for the disease is lost, and by tightening eligibility requirements for the study, it might be more difficult to achieve the desired sample size. Also, if restriction is made too narrow, generalizability to the target population could be jeopardized, especially if features of the disease vary substantially across categories of the variables used for restriction.

Matching

One of the lessons of Chapter 8 is that counterfactual and classical confounding in a closed cohort study are avoided when the confounder has the same distribution in the exposed and unexposed cohorts at the start of follow-up. It is therefore natural when designing a cohort study to arrange for the exposed and unexposed cohorts to have this property. Matching is one of the methods used to accomplish this goal (randomization is another).

There are several types of matching. Here we focus on frequency matching and individual matching. With *frequency matching*, the distribution of the confounder in the exposed cohort is used to guide the recruitment of an unexposed cohort so that the confounder has the same distribution in both cohorts. For this to be possible, it is necessary for the exposed cohort to have been assembled prior to the recruitment of the unexposed cohort, or it must somehow be known in advance of recruiting the unexposed cohort what the confounder distribution of the exposed cohort is likely to be. For example, frequency matching is feasible when the exposed and unexposed subjects are already part of an established database that has the necessary information on confounders of interest.

With *individual matching*, the necessity of having prior information on the distribution of confounders in the exposed cohort is avoided. As each exposed subject is enrolled in the study, one or more unexposed subjects with the same confounder profile as the exposed subject are recruited. Together these individuals form a *matched set*. For example, if age and gender are the confounders of concern, and the exposed subject is a male 20–24 years old, then each of the unexposed subjects in the matched set must also be a male 20–24 years old. Depending on the study design, the number of unexposed individuals matched to each exposed subject might be allowed to vary from matched set to matched set. If there are m unexposed individuals in each matched set, we say that the study has m:1 matching. When $m = 1$, the study is said to be *pair-matched*. There is no reason to distinguish between matched sets that have the same confounder profile; they can be collapsed into a single matched set. For example, suppose there are three matched sets of males 20–24 years old. They can be combined into one matched set with three exposed subjects and as many unexposed subjects as there were in the original matched sets.

The goal of matching in a closed cohort study is to increase validity, as the next example illustrates.

Example: Frequency-matched Closed Cohort Study Based on (Unmatched) Closed Cohort Study #4

Closed cohort study #4 is presented in Tables 9.2(a)–9.2(c) using the usual format. It is readily verified that there is counterfactual and classical confounding. Consider a frequency-matched closed cohort study based on closed cohort study #4 where the exposed cohort for the matched study is a 10% simple random sample of the exposed cohort for the unmatched study. Table 9.3(a) gives the data and risks for the matched study.

The entries in Table 9.3(a) are obtained from Table 9.2(a) as follows. Based on the distribution of subjects at the start of follow-up in the exposed cohort of the unmatched study, there are 800 and 200 subjects at the start of follow-up in the strata of the exposed cohort of the matched study. For the next steps, see the computations below Table 9.3(a). Since the stratum-specific risks in the exposed cohort of the matched study are the same as those in the exposed cohort of the unmatched study, there are 160 and 20 incident cases in the strata of the exposed cohort of the matched study. Based on frequency matching, there are 800 and 200 subjects at the start of follow-up in the strata of the unexposed cohort of the matched study. Since the stratum-specific risks in the unexposed cohort of the matched study are the same as those in the unexposed cohort of the unmatched study, there are 16 and 2 incident cases in the strata of the unexposed cohort of the matched study. This gives Table 9.3(a), and Table 9.3(b) shows the corresponding measures of effect. Comparing the crude and standardized measures of effect, we see that there is no counterfactual confounding in the matched study, and so the crude measures of effect are suitable as overall measures of effect. Thus, there is no need to perform a stratified analysis to control for confounding by F—matching on F is all that is necessary. ◊

Even though a frequency-matched design for a closed cohort study has the potential to enhance validity, such studies are rarely conducted in practice. Perhaps this is because cohort studies often have large sample sizes, which means that matching could be costly and time-consuming. Usually it would be considered more efficient to expend those resources on recruiting additional subjects for an unmatched study, followed by a stratified data analysis. However, matching may be the only viable way to control confounding when the exposed cohort is small or when there are confounders that are difficult to measure. For example, matching on neighborhood of residence might be used to control confounding due to socioeconomic factors that are hard to quantify.

Table 9.2(a) Data and risks for closed cohort study #4.

Stratum	$E=1$			$E=0$		
	Cases	Persons	Risk[a]	Cases	Persons	Risk[a]
$F=1$	1,600	8,000	20.0	40	2,000	2.0
$F=0$	200	2,000	10.0	80	8,000	1.0
Overall	1,800	10,000	18.0	120	10,000	1.2

[a] *per 100.*

Table 9.2(b) Distribution of F at start of follow-up (percent) for closed cohort study #4.

Stratum	$E=1$	$E=0$
$F=1$	80.0	20.0
$F=0$	20.0	80.0

Table 9.2(c) Measures of effect for closed cohort study #4.

Measure of effect	Crude	Stratum-specific		Stand.	M-H
		$F=1$	$F=0$		
Risk ratio	15.0	10.0	10.0	10.0	10.0
Odds ratio	18.1	12.3	11.0	12.0	11.8
Risk difference[a]	16.8	18.0	9.0	16.2	13.5

[a] *per 100.*

Table 9.3(a) Data and risks for frequency-matched closed cohort study based on closed cohort study #4.

Stratum	$E=1$			$E=0$		
	Cases	Persons	Risk[a]	Cases	Persons	Risk[a]
$F=1$	160	800	20.0	16	800	2.0
$F=0$	20	200	10.0	2	200	1.0
Overall	180	1,000	18.0	18	1,000	1.8

[a] *per 100.*

$$160 = (20.0 \times 10^{-2}) \times 800 \qquad 16 = (2.0 \times 10^{-2}) \times 800$$

$$20 = (10.0 \times 10^{-2}) \times 200 \qquad 2 = (1.0 \times 10^{-2}) \times 200$$

Table 9.3(b) Measures of effect for frequency-matched closed cohort study based on closed cohort study #4.

Measure of effect	Crude	Stratum-specific		Stand.	M-H
		$F=1$	$F=0$		
Risk ratio	10.0	10.0	10.0	10.0	10.0
Odds ratio	12.0	12.3	11.0	12.0	12.1
Risk difference[a]	16.2	18.0	9.0	16.2	16.2

[a] *per 100.*

References

1. Greenland S. Variance estimators for attributable fraction estimates consistent in both large strata and sparse data. *Stat Med* 1987;**6**(6):701–8.
2. Greenland S. Bias in methods for deriving standardized morbidity ratio and attributable fraction estimates. *Stat Med* 1984;**3**(2):131–41.
3. Levin ML. The occurrence of lung cancer in man. *Acta Unio Int Contra Cancrum* 1953;**9**(3):531–41.
4. Poole C. A history of the population attributable fraction and related measures. *Ann Epidemiol* 2015;**25**(3):147–54.
5. Miettinen OS. Proportion of disease caused or prevented by a given exposure, trait or intervention. *Am J Epidemiol* 1974;**99**(5):325–32.
6. Walter SD. The estimation and interpretation of attributable risk in health research. *Biometrics* 1976;**32**(4):829–49.
7. Doll R, Peto R, Boreham J, et al. Mortality in relation to smoking: 50 years' observations on male British doctors. *BMJ* 2004 June 26;**328**(7455):1519–28.
8. Cornfield J, Haenszel W, Hammond EC, et al. Smoking and lung cancer: recent evidence and a discussion of some questions. *J Natl Cancer Inst* 1959;**22**(1):173–203.
9. Schlesselman JJ. Assessing effects of confounding variables. *Am J Epidemiol* 1978;**108**(1):3–8.

10. Seigel DG, Greenhouse SW. Validity in estimating relative risk in case-control studies. *J Chron Dis* 1973;**26**(4):219–25.
11. Friedman LM, Furberg CD, DeMets DL, et al. *Fundamentals of clinical trials*. 5th ed. Heidelberg: Springer; 2015.
12. Lash TL, VanderWeele TJ, Haneuse S, et al., editors. *Modern epidemiology*. 4th ed. Philadelphia: Wolters Kluwer; 2021.
13. Collett D. *Modelling survival data in medical research*. 3rd ed. Boca Raton: CRC Press; 2015.
14. Chesnaye NC, Stel VS, Tripepi, et al. An introduction to inverse probability of treatment weighting in observational research. *Clinical Kidney J* 2022;**15**(1):14–20.
15. Hernán MA, Robins JM. *Causal inference: what if*. Boca Raton: Chapman & Hall/CRC; 2020. https://www.hsph.harvard.edu/miguel-hernan/causal-inference-book.

CHAPTER

Bias in Cohort Studies

10

As with virtually all epidemiologic study designs, cohort studies are vulnerable to a seemingly endless array of biases, many of which have received their own name. Here we focus on a few key biases that are of particular relevance to the cohort approach.

10.1 Selection Bias

The *healthy worker effect* is a form of selection bias that arises in occupational studies. Workers employed in physically demanding jobs tend to be more vigorous than the average person, at least when they first enter the workforce. As they continue in their occupation, their health gradually approaches population norms. This means that in occupational studies of mortality comparing workers to the general population, the mortality risk due to an occupational exposure will tend to be underestimated. A classic example is a cohort study in Great Britain of 7,561 men engaged in fabricating vinyl chloride monomer at some time during 1940–1974.[1] The standardized mortality ratio for all causes of death after 35 years of follow-up was 0.75, varying from a low of 0.41 for the first five years to a high of 1.07 for the last five years. The healthy worker effect can be mitigated by delaying entry of workers into the occupational cohort until they have passed through a period of relative good health.

Although the healthy worker effect is usually thought of as a selection bias, it has an alternative interpretation as a type of confounding. According to this point of view, the problem lies not with the workers being unusually healthy, but rather with the general population being chosen as the comparison group. A more suitable choice would be, for example, another occupational cohort working in an equally demanding job. Alternatively, if information is available on the overall health of the workers and the general population, stratification or regression analysis can be used to control for the confounding effects of differences in overall health in the two study groups.

Lost to follow-up bias is a type of selection bias due to differing patterns of loss to follow-up in the exposed and unexposed cohorts. The implications of this form of bias were discussed in Section 4.2 in connection with noninformative censoring in open cohort studies. There is little that can be done to mitigate this type of bias other than trying to keep losses to follow-up to a minimum.

Epidemiologic Methods. https://doi.org/10.1016/B978-0-44-318780-3.00016-6
Copyright © 2023 Elsevier Inc. All rights reserved.

10.2 Information Bias

Detection bias (or *ascertainment bias*) is a type of information bias that occurs when the likelihood of detecting an outcome of interest differs in the exposed and unexposed cohorts. For example, consider a cohort study of estrogen therapy as a risk factor for endometrial cancer. Estrogen causes vaginal bleeding, which is a symptom of endometrial cancer and a reason for seeking medical care. This could lead to endometrial cancer being detected in women taking estrogen when it otherwise would not have been diagnosed as early.

Nondifferential Misclassification of Case Status

Misclassification of case status in prevalence studies was discussed in Section 3.2. Let us now examine misclassification of case status in closed cohort studies. We say that the misclassification of case status is *nondifferential* if it does not depend on exposure status, and *differential* otherwise.

Consider a closed cohort study in which there is nondifferential misclassification of case status. By definition, the "test" used to determine case status has the same sensitivity (S) in the exposed and unexposed cohorts, and likewise for the specificity (T). Our aim is to measure the impact of misclassification on the risk ratio (RR), odds ratio (OR) and risk difference (RD). Keeping with the star notation of Section 3.2, we denote their misclassified counterparts by $RR^\star$, $OR^\star$ and $RD^\star$. It is shown in Appendix I that

$$RR^\star = \frac{1 - T + (S + T - 1)R_1}{1 - T + (S + T - 1)R_0}$$

$$OR^\star = \frac{[1 - T + (S + T - 1)R_1][T - (S + T - 1)R_0]}{[1 - T + (S + T - 1)R_0][T - (S + T - 1)R_1]} \tag{10.1}$$

$$RD^\star = (S + T - 1)RD\,.$$

Let us assume that $0 < S < 1$ and $0 < T < 1$ and $S + T > 1$, and also that $0 < R_0 < 1$ and $0 < R_1 < 1$. These conditions are almost always satisfied in practice. It can be shown using (10.1) that $RR^\star = 1$ if and only if $RR = 1$, and $RD^\star = 0$ if and only if $RD = 0$. Less obvious but true is that $OR^\star = 1$ if and only if $OR = 1$, which is demonstrated in Appendix I. Suppose none of RR, OR and RD equals its null value, or equivalently that $R_1 \neq R_0$. It is shown in Appendix I that for the risk ratio,

$$RR > RR^\star > 1 \quad \text{or} \quad RR < RR^\star < 1\,, \tag{10.2}$$

for the odds ratio,

$$OR > OR^\star > 1 \quad \text{or} \quad OR < OR^\star < 1\,, \tag{10.3}$$

and for the risk difference,

$$RD > RD^\star > 0 \quad \text{or} \quad RD < RD^\star < 0\,. \tag{10.4}$$

Thus, nondifferential misclassification of case status biases each of the measures of effect toward its null value.[2,3] When the misclassification of case status is differential, the direction of bias can be either toward or away from the null value.

Example: Nondifferential Misclassification of Case Status in Closed Cohort Study #5

Table 10.1(a) gives the data and risks for closed cohort study #5 (using the layout of Table 4.1). The crude measures of effect are $RR = 2.00$, $OR = 2.25$ and $RD = 0.10$. Let us examine the effects of nondifferential misclassification of case status when the sensitivity and specificity are 80% and 90%, respectively, where we note that $80\% + 90\% > 100\%$. Tables 10.1(b) and 10.1(c) give the misclassified data for the exposed and unexposed cohorts, respectively, where computations are performed as in Table 3.2. The far-right columns of Tables 10.1(b) and 10.1(c) provide the first two rows of Table 10.1(d), which gives the misclassified data and risks based on closed cohort study #5. The misclassified crude measures of effect are $RR^\star = 1.41$, $OR^\star = 1.54$ and $RD^\star = 0.07$, each of which is biased toward the null value, as expected. ◊

Table 10.1(a) Data and risks for closed cohort study #5.

Exposure	$D=1$	$D=0$	Persons	Risk
$E=1$	20	80	100	0.20
$E=0$	10	90	100	0.10
Overall	30	170	200	0.15

Table 10.1(b) Misclassified data for exposed subjects in closed cohort study #5: sensitivity 80% and specificity 90%.

	$D=1$	$D=0$	Total
$D^\star=1$	16 = 80% × 20	8	24
$D^\star=0$	4	72 = 90% × 80	76
Total	20	80	100

Table 10.1(c) Misclassified data for unexposed subjects in closed cohort study #5: sensitivity 80% and specificity 90%.

	$D=1$	$D=0$	Total
$D^\star=1$	8 = 80% × 10	9	17
$D^\star=0$	2	81 = 90% × 90	83
Total	10	90	100

Table 10.1(d) Misclassified data and risks for closed cohort study #5: sensitivity 80% and specificity 90%.

Exposure	$D^\star = 1$	$D^\star = 0$	Persons	Risk
$E = 1$	24	76	100	0.24
$E = 0$	17	83	100	0.17
Overall	41	159	200	0.21

More generally, for a closed cohort study with nondifferential misclassification of case status, Fig. 10.1 shows the ratio $OR^\star/OR$ as a function of R_0 for $OR = 2, 5, 10$, where the sensitivity and specificity are fixed at 80% and 90%, respectively. The formula for $OR^\star/OR$ used in Fig. 10.1 is given in Appendix I. Since each of the selected values of OR is greater than 1, it follows from (10.3) that $OR^\star/OR < 1$ for all values of R_0, and this is reflected in the graph. For a given value of R_0, the bias due to nondifferential misclassification increases as OR increases.

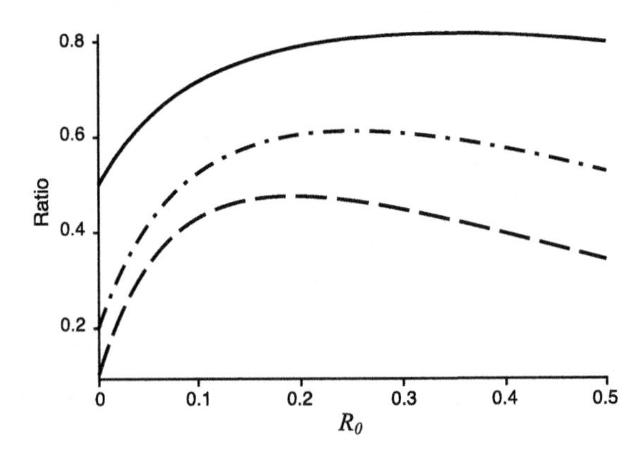

FIGURE 10.1

$OR^\star/OR$ as a function of R_0 for a closed cohort study with nondifferential misclassification of case status: sensitivity 80% and specificity 90% ($OR = 2$: *solid*; $OR = 5$: *dashdot*; $OR = 10$: *dash*).

References

1. Fox AJ, Collier PF. Mortality experience of workers exposed to vinyl chloride monomer in the manufacture of polyvinyl chloride in Great Britain. *Br J Ind Med* 1977;**34**(1):1–10.
2. Copeland KT, Checkoway H, McMichael AJ, et al. Bias due to misclassification in the estimation of relative risk. *Am J Epidemiol* 1977;**105**(5):488–95.
3. Gullen WH, Bearman JE, Johnson EA. Effects of misclassification in epidemiologic studies. *Public Health Rep* 1968;**83**(11):914–8.

CHAPTER

11 Randomized Trials

A *randomized trial* (or *randomized controlled trial*) is a type of cohort study that comes close to achieving the counterfactual ideal. For that reason, findings from a randomized trial are usually given the highest priority in the hierarchy of evidence from epidemiologic studies. The methodology of randomized trials is ideally suited to the evaluation of health care innovations in such areas as pharmaceuticals, surgical procedures and medical devices. Not every epidemiologic question regarding health care is amenable to a randomized trial. The reason often has to do with ethics, which is discussed below.

Since the majority of randomized trials are conducted in clinical or public health settings, the terminology used is somewhat different from that employed in observational cohort studies. Specifically, we speak of *intervention* and *control* groups (rather than exposed and unexposed cohorts), *treatments* (rather than exposures), *treatment groups* (referring to the intervention and control groups), and *outcomes* of interest (rather than diseases of interest).

The following two terms appear frequently in the literature on randomized trials. The *efficacy* of a treatment has to do with how well it works under ideal circumstances, whereas the *effectiveness* of a treatment is concerned with its performance under real-world conditions. For example, a government agency responsible for granting marketing approval is usually preoccupied with efficacy, whereas an organization involved in funding health care might be more focused on effectiveness. An individual who has been prescribed the treatment and is prepared to make use of it in an optimal fashion will be interested in its efficacy rather than its effectiveness.

We consider three types of randomized trials—randomized clinical trials, randomized field trials and community intervention trials.

11.1 Randomized Clinical Trials

Most randomized clinical trials have what is called a parallel design. In a *parallel trial*, subjects are randomly assigned to either the intervention or control group and then followed for the occurrence of one or more outcomes of interest. Less commonly seen are the factorial and crossover designs. In a *factorial trial*, each subject is enrolled simultaneously in two or more overlapping parallel trials; while in a *crossover trial*, the same individuals serve alternately as intervention and control subjects. Most of the remarks to follow on the parallel design can be applied directly or readily

Epidemiologic Methods. https://doi.org/10.1016/B978-0-44-318780-3.00017-8
Copyright © 2023 Elsevier Inc. All rights reserved.

adapted to the factorial and crossover designs. For concreteness, we focus on randomized clinical trials of drugs.

Parallel Design

Phase I–IV Clinical Trials

The process of creating and bringing a new drug to market is a complex and costly undertaking. Once the drug has been synthesized, animal studies of pharmacology and toxicology are conducted. If these investigations go well, human studies in the form of phase I–IV clinical trials follow.

A *phase I clinical trial* is conducted on healthy volunteers or, in the case of a drug with potentially significant toxicity, on patients with advanced disease who are willing to take a chance on a promising but untried treatment. The goal is to assess safety and tolerability, establish dosing regimens, and investigate the pharmacokinetics (absorption, distribution, metabolism, excretion) and pharmacodynamics (dose-response relationship) of the drug. A *phase II clinical trial* is an exploratory therapeutic study conducted on patients with the disease, usually selected according to narrow criteria to ensure a relatively homogeneous study group. The aim is to determine if the drug has biologic activity, refine dosing regimens, and further assess safety and tolerability. A *phase III clinical* trial is a definitive therapeutic study conducted on patients with the disease. The expression "phase III clinical trial" is essentially synonymous with "randomized clinical trial". The goal is usually to evaluate the efficacy of the drug for the purpose of seeking marketing approval from a government agency. A *phase IV clinical trial* is an observational study conducted to evaluate the effectiveness of the drug after marketing approval has been received. This type of study might be undertaken, for example, as part of postmarketing surveillance by the drug manufacturer.

Ethics

It goes without saying that any epidemiologic study must be conducted according to high ethical standards.[1,2(p.25–48)] The responsibility for meeting this obligation is heightened in clinical trials because, unlike in observational studies, an intervention is being offered that may have unknown risks. The principle of *equipoise* provides an ethical basis for randomized clinical trials.[3] It refers to uncertainty in the expert medical community (not just on the part of the clinical personnel involved in a trial) over whether a new treatment might be better than available treatments. If there is consensus on which treatment is more beneficial, the clinical trial should not be conducted.

To be considered ethical, a clinical trial must be methodologically rigorous and have a clear scientific objective, an adequate sample size, a valid mechanism for data collection, and an appropriate plan for data analysis. When designing a randomized clinical trial, there may be several treatment options for controls, such as one of the currently accepted treatments, a placebo, or no treatment at all. In a drug trial, a *placebo* is a pharmacologically inactive agent that is designed to resemble the active drug in terms of appearance and side effects. In certain settings, a placebo can have an effect on outcome, referred to as a *placebo effect*. This is commonly seen in randomized clinical trials of psychiatric medications, for example.

Participation in a clinical trial must be voluntary and accompanied by an *informed consent* that covers at least the following areas: aims of the study; procedures to be followed; sources of funding; possible conflicts of interest on the part of sponsors and investigators; anticipated benefits and potential risks of the study to participants; how confidentiality will be protected; whom to contact with questions; a statement that participation is voluntary, that there is no penalty for not participating, and that there will be no reprisals for withdrawing. Approval for a clinical trial needs to be obtained from an *institutional research board (IRB)* or similar body at each participating institution. The IRB should include bioethicists, clinicians, epidemiologists, pharmacologists, members of the lay public, and possibly others.

Randomization

Randomization (or *random allocation*), the process of randomly assigning subjects to intervention and control groups, is the defining feature of randomized clinical trials.[2(p.123–145),4] Not all clinical trials are randomized. For example, phase I and phase II clinical trials discussed above are not randomized, nor was the Lind scurvy trial discussed in Section 1.1.

The goal of randomization is to reduce selection bias and confounding. Selection bias can occur if assignment to a treatment group is left to the discretion of the investigators. For example, physicians involved in the trial might want patients believed to have a good prognosis to receive the intervention. This would result in an intervention group heavily weighted toward patients likely to do well. Randomization avoids this type of problem by removing choice from the process of group allocation, as described below.

Confounding occurs when the intervention and control groups have different confounder profiles. Randomization has the potential to eliminate, or at least mitigate, this problem. To the extent that randomization produces treatment groups with similar confounder profiles, the groups are said to be more or less *balanced*. An important feature of randomization is that it affects both measured and unmeasured confounders. However, because randomization is based on chance, there is no guarantee that perfect or even acceptable balance will be achieved in any given randomized clinical trial. Indeed, randomization might yield a very skewed distribution of confounders. Perfect balance is an ideal that is achieved only "on average", meaning that if randomization were to be repeated many times, the trend would be in the direction of perfect balance. For a given study, the larger the sample size, the greater the prospects of achieving acceptable balance.

There are several types of randomization, three of which are discussed here—simple randomization, blocked randomization and stratified randomization.

Simple randomization can be accomplished by flipping a coin, although more sophisticated procedures such as those using computer-based random number generators are often employed in practice. A virtue of simple randomization is that the treatment group to which a subject is assigned cannot be predicted in advance. A disadvantage is that the number of subjects in the treatment groups might be grossly unequal. For a randomized clinical trial with 20 subjects, the probability of a 60:40% split or one more extreme (that is, 12 or more subjects in one of the groups) is 50.3%, whereas for a clinical trial with 100 subjects, the probability of a 60:40% split or one more extreme (that is, 60 or more subjects in one of the groups) is only 5.7%. Such unequal group allocation does not invalidate statistical tests but it does reduce statistical power. Although it is not clear from the published report, the MRC streptomycin trial discussed in Section 1.8 may have used simple randomization. This would account for the slight difference in numbers of patients in the treatment groups: 55 treated with streptomycin plus bed rest, and 52 treated with bed rest only.

Blocked randomization (or *blocking*) involves assigning subjects to treatment groups in what are called *blocks*, each of which is constructed with a given intervention-to-control ratio. When the ratio is 1:1, blocking ensures that there are equal numbers of subjects in the intervention (I) and control (C) groups. To illustrate, for a design with a 1:1 ratio and a block size of four, there are six possible blocks: IICC, CCII, ICIC, CICI, ICCI, CIIC. If, for example, the IICC block is used, the first two of the next four subjects are assigned to the intervention group, and the next two subjects to the control group. If the blocks are used in a fixed order, group allocation may be predictable. For example, if it is known that the IICC block is being used and three subjects have already been assigned, the next subject will necessarily be placed in the control group. The choice of blocks can be made random to increase unpredictability.

Simple randomization and blocked randomization ensure that confounder profiles are similar across treatment groups "on average", but not in any given randomized clinical trial. *Stratified randomization* involves stratifying subjects according to one or more (measured) confounders and then performing blocked randomization within each stratum. This guarantees balance on the chosen confounders.

Example: University Group Diabetes Program

The University Group Diabetes Program (UGDP) was a multicenter, double-blind, placebo-controlled, parallel randomized clinical trial conducted to investigate whether controlling blood sugar with one of several treatments reduces the development of vascular complications of diabetes.[5,6] Individuals with at most a one-year history of diabetes were recruited during 1961–1966 and randomly assigned to one of five treatment groups using stratified randomization, with stratification by study center and randomization in blocks of length 16 based on type of treatment. Here we focus on the comparison of the oral medication tolbutamide to placebo, with death from any cause as the outcome of interest. There were 204 subjects in the tolbutamide group and 205 in the placebo group. Tables 11.1(a)–11.1(c) give the data and mortality risks, age distribution at the start of treatment, and measures of effect, respectively.

We see from Table 11.1(b) that randomization did not produce balance on age group, with the tolbutamide group somewhat older than the placebo group. The crude

Table 11.1(a) Data and risks for UGDP study.

Age group	Tolbutamide			Placebo		
	Deaths	Persons	Risk[a]	Deaths	Persons	Risk[a]
55+	22	98	22.4	16	85	18.8
<55	8	106	7.5	5	120	4.2
Overall	30	204	14.7	21	205	10.2

[a] *per 100.*
Data from Reference 6.

Table 11.1(b) Age distribution at start of follow-up (percent) in UGDP study.

Age group	Tolbutamide	Placebo
55+	48.0	41.5
<55	52.0	58.5

Data from Reference 6.

Table 11.1(c) Measures of effect for UGDP study: tolbutamide is exposure and age group is stratification variable.

Measure of effect	Crude	Stratum-specific		Stand.	M-H
		55+	<55		
Risk ratio	1.44	1.19	1.81	1.31	1.33
Odds ratio	1.51	1.25	1.88	1.37	1.40
Risk difference[a]	4.46	3.63	3.38	3.50	3.49

[a] *per 100.*
Data from Reference 6.

mortality risk is greater in the tolbutamide group (14.7%) than in the placebo group (10.2%), but there is evidence of moderate counterfactual confounding due to age, especially apparent with the risk difference. Based on the standardized measures of effect, tolbutamide appears to result in increased mortality compared to placebo, a surprising finding that provoked much controversy. ◊

Blinding

Information bias can be introduced into a randomized clinical trial if assignment to treatment group is known to subjects or investigators, or both. For example, subjects who know they are in the intervention group might consciously or subconsciously try to please those who are monitoring their health by minimizing symptoms. Procedures instituted to ensure that group allocation is hidden are referred to as *blinding* (or *masking*).[2(p.147–163), 7] The goal of blinding is to reduce information bias, especially *response bias* on the part of subjects, and *ascertainment bias* (or *observer bias*) on the part of the investigators responsible for assessing the health status of subjects.

In an *unblinded trial* (or *open trial*), both the subjects and the investigators know the treatment group to which subjects have been assigned. Sometimes this is unavoidable, as in a trial that compares a surgical to a medical treatment. An unblinded trial has the attractions that it is logistically easier to conduct than one that is blinded, and those responsible for the health care of patients may be more comfortable making clinical decisions if group allocation is known. The main disadvantages are the increased likelihood of response bias and ascertainment bias. In a *single-blind* trial, the investigators but not the subjects know the treatment group to which subjects have been assigned. In a *double-blind* trial, neither the investigators nor the subjects know the treatment group to which subjects have been allocated.

Example: Vitamin C Trial

A double-blind, placebo-controlled, parallel randomized clinical trial was conducted to investigate whether vitamin C helps to prevent the common cold and reduce symptoms if one occurs.[8,9] A total of 311 subjects meeting eligibility criteria were randomly assigned using simple randomization to one of four treatment groups: a constant daily dose of vitamin C; a constant daily dose of placebo; a constant daily dose of vitamin C plus supplemental vitamin C if cold symptoms developed; a constant daily dose of placebo plus supplemental placebo if cold symptoms developed. The outcomes of interest were: experiencing a cold and the subjective symptom count if one occurred. The planned follow-up was one year, but the trial was stopped after nine months due to dropouts, at which point there were 190 participants. Among subjects who claimed not to know their treatment, those taking vitamin C and those taking placebo had a similar duration of symptoms. However, among subjects who claimed to know their treatment, those taking vitamin C had a shorter duration of symptoms than those taking placebo. ◊

Adherence

Adherence (or *compliance*) is concerned with the extent to which subjects follow the treatment regimen to which they have been assigned. There are a number of factors that influence adherence.[2(p.297–318)] Not surprisingly, the shorter the trial and the simpler the treatment regimen, the better the adherence. Trials where subjects are under close supervision, such as in a hospital, tend to have fewer problems with adherence. On the other hand, trials based in the community, especially those that require a change in personal behaviors, are more likely to experience low adherence. Using a *run-in period* prior to randomization, where potential subjects have an opportunity to experience a treatment regimen before being officially enrolled in the trial, helps to exclude those likely to have poor adherence.

Intention to Treat and Per Protocol Analyses

Nonadherence has implications for the way in which data from a randomized clinical trial are analyzed. Here we consider two fundamental approaches—*intention to treat* and *per protocol*.[10,11] Broadly speaking, an intention to treat analysis is concerned with the effect of the treatment assigned, whereas a per protocol analysis is concerned with the effect of the treatment received.

In an intention to treat analysis, subjects are included in the treatment group to which they were assigned, regardless of whether they adhered to their treatment regimen or not. This means, for example, that a subject who was assigned to the intervention drug but chose not to take it would nevertheless be included in the intervention group for the purposes of an intention to treat analysis. Of course, in the day to day world, patients do not always follow treatment recommendations. Thus, an intention to treat analysis is concerned with effectiveness. The advantages of an intention to treat analysis are that it retains the original sample size and preserves randomization. However, the mismatch between treatment assigned and treatment received can result in a biased measure of effect, as we now illustrate.

Consider a placebo-controlled randomized clinical trial of a drug where the placebo has no effect on outcome, and patients in the placebo group do not have access to the drug. Suppose there is nonadherence in the drug group and that an intention to treat analysis is conducted. If the drug has no effect on outcome, the estimated measure of effect will (on average) equal its null value, regardless of the degree of nonadherence. However, if the drug has an effect on outcome, the estimated measure of effect will (on average) be biased toward the null value because of "dilution" by nonadherent subjects in the drug group.[11]

In a *per protocol* analysis, subjects are included in the data analysis only if they adhered to their assigned treatment regimen. Thus, this approach is focused on efficacy. A per protocol analysis avoids the mismatch between treatment assigned and treatment received seen with an intention to treat analysis, but has the disadvantages of a reduced sample size and, most importantly, the possibility of selection bias due to restricting the data analysis to only those subjects who were adherent.

Example: Coronary Drug Project

The Coronary Drug Project was a multicenter, double-blind, placebo-controlled, parallel randomized clinical trial conducted to investigate, among other things, whether the lipid-lowering drug clofibrate reduces death from any cause in individuals with coronary heart disease.[12] To be eligible, subjects needed to be male, 30–64 years old, and have electrocardiographic evidence of a heart attack that had occurred not less than three months previously. Subjects were recruited during 1966–1969, with follow-up for the clofibrate and placebo groups until 1974. A total of 1,103 individuals were assigned to clofibrate and 2,789 to placebo. Subjects were said to be adherent if at least 80% of capsules were taken, and nonadherent otherwise.

Table 11.2 gives the number of subjects in the clofibrate and placebo groups for whom adherence data were available, as well as 5-year mortality risks for all causes of death, stratified by adherence status. The last row gives the (crude) intention to treat analysis and shows a slight reduction in mortality risk from clofibrate (18.2%) compared to placebo (19.4%). The row for adherent subjects gives the (crude) per

Table 11.2 Adherence data and mortality risks for Coronary Drug Project.

Adherent	Clofibrate		Placebo	
	Persons	Risk[a]	Persons	Risk[a]
Yes	708	15.0	1,813	15.1
No	357	24.6	882	28.2
Overall	1,065	18.2	2,695	19.4

[a] *per 100.*
Data from Reference 12.

protocol analysis and shows essentially no reduction in mortality risk from clofibrate (15.0%) compared to placebo (15.1%).

Focusing on the clofibrate group, the mortality risk difference of 9.6% in adherent subjects compared to nonadherent subjects suggests that clofibrate is responsible for a reduction in mortality. However, there is a corresponding mortality risk difference of 13.1% in the placebo group. It is inconceivable that the latter mortality reduction is due entirely to a placebo effect. The Coronary Drug Program investigators conjectured that the mortality reduction in the placebo group was due to differences in characteristics of those who were adherent compared to those who were not. They performed a multiple linear regression of 5-year mortality risk on adherence in placebo subjects that involved 40 baseline variables. The model-based mortality risk difference was 9.4%. This finding has been widely accepted as empirical evidence that it is not possible to adjust for differences in adherence in randomized clinical trials. ◊

The difficulty adjusting for selection bias in a per protocol analysis, as exemplified by the Coronary Drug Program, has led to a preference for the intention to treat approach to randomized clinical trials. However, this stance appears to be softening. A reanalysis of Coronary Drug Program data using statistical methods developed since that study was conducted shows that it may be possible to account for most of the effects of nonadherence in a per protocol analysis.[13,14] In its 2010 statement on guidelines for reporting parallel randomized clinical trials, the authoritative Consolidated Standards of Reporting Trials (CONSORT) states that an intention to treat analysis is "generally favored" but that a per protocol analysis "may be appropriate in some settings".[15,16]

Factorial Design

In a randomized clinical trial with a 2 × 2 *factorial design*, two treatments (with their corresponding outcomes) are investigated in a single study. Let us denote the treatments by A and B. Each subject in such a trial is randomly assigned to receive treatment A or not, and simultaneously randomized to receive treatment B or not. This gives four treatment groups: A and B; A but not B; B but not A; neither A nor B. The analysis and interpretation of trial data is made more complicated when the treatments exert a joint effect on one or both of the outcomes.

Example: Women's Health Initiative

Observational studies conducted prior to 2000 showed that estrogen therapy with or without progestin therapy in postmenopausal women reduces the risk of coronary heart disease and increases the risk of breast cancer.[17–20] This led to recommendations supporting the use of hormone replacement therapy (HRT) for cardiovascular protection. The Women's Health Initiative (WHI) was a multicenter, double-blind, placebo-controlled, $3 \times 2 \times 2$ partial factorial randomized clinical trial that investigated the health effects of HRT and dietary modification.[21] Here we focus on the HRT component. A total of 16,608 postmenopausal women 50–79 years old with an intact uterus were recruited from 40 clinical centers in the United States during 1993–1998. Using stratified randomization, with stratification by study center and age group, subjects were randomized to receive daily estrogen plus progestin ($n = 8,506$) or placebo ($n = 8,102$).

The primary outcomes of interest were coronary heart disease and invasive breast cancer. The planned follow-up was eight and a half years, but the trial was stopped after just over five years due to excess invasive breast cancer in the HRT group. Based on an intention to treat analysis, the hazard ratios were: coronary heart disease 1.29, invasive breast cancer 1.26, and death from all causes 0.98. The contradictory findings from earlier observational studies and the WHI regarding the risk of coronary heart disease associated with HRT in postmenopausal women engendered a protracted debate that seems finally to have been resolved.[22] ◊

Crossover Design

In a randomized clinical trial with a *crossover design*, each subject is randomly assigned to receive either the intervention or control treatment, and after an appropriate *wash-out period*, during which the treatment is discontinued, the subject is switched to the alternate treatment.[23] In this way, subjects act as their own controls, making this type of study the closest approximation to the counterfactual ideal of any epidemiologic design. In order for a crossover trial to be successful, it is necessary for the intervention and control treatments to be short-lasting with no carryover effects from one treatment to the other, and for the disease to be stable over time and promptly revert to its baseline status once the intervention and control treatments are withdrawn. These conditions ensure that a change in health status can be attributed to the treatment being taken at the time, not to the previous treatment or to inherent changes in the disease. For example, these criteria are satisfied by short-acting analgesics taken for chronic pain conditions such as arthritis.

Example: Migraine Trial

A double-blind, placebo-controlled, crossover randomized clinical trial was conducted to investigate whether daily use of the antihypertensive medication candesartan helps to prevent migraine headaches.[24] During January 2001 to February 2002, 60 individuals 18–65 years old with a history of two to six migraine headaches per month for at least a year were recruited through newspaper advertisements and from an outpatient clinic. A placebo run-in period of four weeks was followed by two 12-week treatment periods separated by a 4-week washout period. Thirty patients were randomly assigned to receive candesartan in the first treatment period, followed by placebo in the second treatment period. The remaining 30 patients received the placebo followed by candesartan. An intention to treat analysis of data recorded in headache diaries showed that the mean number of days with migraine headache was 13.6 when subjects were taking candesartan and 18.5 when they were taking placebo. ◊

11.2 Randomized Field Trials

A randomized field trial is similar to a randomized clinical trial with a parallel design, except that it is conducted in the general population and has the goal of evaluating whether a public health intervention prevents disease in at-risk individuals. Such trials usually require a large sample size, making them logistically challenging and costly.

To make matters concrete, suppose the randomized field trial is designed to evaluate a new vaccine. We assume that there is no counterfactual confounding, which is a reasonable assumption in a large randomized field trial. The benefit of vaccination can then be quantified using the crude exposed attributable fraction, defined by (9.5) and referred to in the present context as the *vaccine efficacy*. We take the unvaccinated group to be the "exposed" cohort, which we label by U, and the vaccinated group to be the "unexposed" cohort, which we labeled by V. The risk ratio is $R_{\mathrm{U}}/R_{\mathrm{V}}$, and so (9.5) becomes[25]

$$\text{vaccine efficacy} = \frac{R_{\mathrm{U}} - R_{\mathrm{V}}}{R_{\mathrm{U}}}\,.$$

By definition, the vaccine efficacy is the proportion of infections in the unvaccinated group that would have been prevented if the unvaccinated group had been vaccinated.

Example: Salk Polio Vaccine Trial

For the placebo control component of the Salk polio vaccine trial, we have from Section 1.9 that

$$\text{vaccine efficacy} = \frac{(57.2 \times 10^{-5}) - (16.4 \times 10^{-5})}{57.2 \times 10^{-5}} = 71.3\%\,.$$

The public health interpretation is that about seven out of ten cases of paralytic polio would be prevented by a vaccination program in which children completed a full series of injections. ◊

11.3 Community Intervention Trials

Not all public health interventions can be evaluated using a randomized field trial. Consider the Newburgh-Kingston fluoride trial discussed in Section 1.10. In theory, it might have been possible to recruit school children from one or more communities and randomize them individually to receive either fluoridated or unfluoridated drinking water for several years, but the logistical problems involved in such an undertaking are obvious. The investigators chose a pragmatic alternative: select two communities that are broadly similar in size and sociodemographic makeup; conduct baseline dental assessments of school children in both communities; select one of the communities to have its drinking water supply fluoridated, leaving the water supply of the other community untouched; after ten years conduct dental assessments on a representative sample of school children in both communities; and compare. Although the dental assessments were performed on an individual basis, it was overall dental health in the two communities, as measured by mean DMF counts, that was used in the comparison.

The Newburgh-Kingston fluoride trial is an example of what is called a *community intervention trial* (CIT).[26,27] The key feature of a CIT is that the public health intervention is conducted at the community level rather than at the individual level. After selecting the communities for study and deciding which of them will receive the intervention, the next step is usually to conduct a survey in each of the communities to obtain baseline measurements. As the trial progresses, further community measurements are taken on a periodic basis. This might take the form of a series of surveys in each community, or the follow-up of a cohort identified at baseline, or both. When there are enough communities involved, the design of the trial can be strengthened by including randomization. The process of randomizing communities proceeds in much the same way as it does for individuals in randomized clinical trials. With only two communities involved, the Newburgh-Kingston fluoride trial would not have benefited from this added feature.

Community trials are typically conducted over a period of several years, which may be sufficient time for changes in background health behavior to have an effect on both the intervention and control communities. This underscores the important role that control communities play in CITs by making it possible to separate the impact of the intervention from the "noise" of temporal change.

Example: COMMIT Trial

The Community Intervention Trial for Smoking Cessation (COMMIT) was a randomized CIT conducted to evaluate the effect of a community-wide anti-smoking campaign.[28,29] The trial involved 11 pairs of communities, ten in the United States and one in Canada, matched on geographic location, size and sociodemographic factors. The sizes of the communities varied from approximately 50,000 to 250,000 residents. During early 1988, a baseline prevalence study was conducted in each community to determine the point prevalence of smoking and to recruit a cohort of smokers for follow-up. One community in each pair was randomly chosen to receive a multifaceted public health campaign during January 1989 to December 1992 directed at smoking cessation, while the other community was only observed. At the close of the intervention period, a second prevalence study of smoking was conducted. The antismoking campaign resulted in no meaningful change in the point prevalence of smoking and no difference in the smoking cessation rates of heavy smokers, but it did have a modest effect on smoking cessation in light-to-moderate smokers. ◊

References

1. Emanuel EJ, Wendler D, Grady C. What makes clinical research ethical? *JAMA* 2000;**283**(20):2701–11.
2. Friedman LM, Furberg CD, DeMets DL, et al. *Fundamentals of clinical trials*. 5th ed. Heidelberg: Springer; 2015.
3. Freedman B. Equipoise and the ethics of clinical research. *N Engl J Med* 1987;**317**(3):141–5.
4. Schulz KF, Grimes DA. Generation of allocation sequences in randomised trials: chance, not choice. *Lancet* 2002 Feb 9;**359**(9305):515–9.
5. Klimt CR, Knatterud GL, Meinert CL, et al. for the University Group Diabetes Program. A study of the effects of hypoglycemic agents on vascular complications in patients with adult-onset diabetes, I: design, methods and baseline results. *Diabetes* 1970;**19**(suppl 2):747–88.
6. Meinert CL, Knatterud GL, Prout TE, et al. for the University Group Diabetes Program. A study of the effects of hypoglycemic agents on vascular complications in patients with adult-onset diabetes, II: mortality results. *Diabetes* 1970;**19**(suppl 2):789–830.
7. Schulz KF, Grimes DA. Blinding in randomised trials: hiding who got what. *Lancet* 2002 Feb 23;**359**(9307):696–700.
8. Karlowski TR, Chalmers TC, Frenkel LD, et al. Ascorbic acid for the common cold: a prophylactic and therapeutic trial. *JAMA* 1975;**231**(10):1038–42.
9. Lewis TL, Karlowski TR, Kapikian AZ, et al. A controlled clinical trial of ascorbic acid for the common cold. *Ann N Y Acad Sci* 1975;**258**:505–12.
10. Tripepi G, Chesnaye NC, Dekker FW, et al. Intention to treat and per protocol analysis in clinical trials. *Nephrology* 2020;**25**(7):513–7.
11. Hernán MA, Hernández-Díaz S. Beyond the intention-to-treat in comparative effectiveness research. *Clin Trials* 2012;**9**(1):48–55.
12. The Coronary Drug Project Research Group. Influence of adherence to treatment and response of cholesterol on mortality in the Coronary Drug Project. *N Engl J Med* 1980;**303**(18):1038–41.
13. Murray EJ, Hernán MA. Adherence adjustment in the Coronary Drug Project: a call for better per-protocol effect estimates in randomized trials. *Clin Trials* 2016;**13**(4):372–8.
14. Murray EJ, Hernán MA. Improved adherence adjustment in the Coronary Drug Project. *Trials* 2018;**19**(1):158. https://doi.org/10.1186/s13063-018-2519-5 [published 5 Mar 2018].
15. Schulz KF, Altman DG, Moher D, for the CONSORT Group. CONSORT 2010 statement: updated guidelines for reporting parallel group randomised trials. *BMJ* 2010;**340**:c332. https://doi.org/10.1136/bmj.c332 [published 23 Mar 2010].

16. Moher D, Hopewell S, Schulz KF, et al. CONSORT 2010 explanation and elaboration: updated guidelines for reporting parallel group randomised trials. *BMJ* 2010;**340**:c869. https://doi.org/10.1136/bmj.c869 [published 23 Mar 2010].
17. Stampfer MJ, Colditz GA. Estrogen replacement therapy and coronary heart disease: a quantitative assessment of the epidemiologic evidence. *Prev Med* 1991;**20**(1):47–63.
18. Grady D, Rubin SM, Petitti DB, et al. Hormone therapy to prevent disease and prolong life in postmenopausal women. *Ann Intern Med* 1992;**117**(12):1016–37.
19. Collaborative Group on Hormonal Factors in Breast Cancer. Breast cancer and hormone replacement therapy: collaborative reanalysis of data from 51 epidemiological studies of 52 705 women with breast cancer and 108 411 women without breast cancer. *Lancet* 1997 Oct 11;**350**(9084):1047–59.
20. Barrett-Connor E, Grady D. Hormone replacement therapy, heart disease, and other considerations. *Annu Rev Public Health* 1998;**19**:55–72.
21. Writing Group for the Women's Health Initiative Investigators. Risks and benefits of estrogen plus progestin in healthy postmenopausal women: principal results from the Women's Health Initiative randomized controlled trial. *JAMA* 2002;**288**(3):321–33.
22. Vandenbroucke JP. The HRT controversy: observational studies and RCTs fall in line. *Lancet* 2009 Apr 11;**373**(9671):1233–5.
23. Louis TA, Lavori PW, Bailar III JC, et al. Crossover and self-controlled designs in clinical research. *N Engl J Med* 1984;**310**(1):24–31.
24. Tronvik E, Stovner LJ, Helde G, et al. Prophylactic treatment of migraine with an angiotensin II receptor blocker: a randomized controlled trial. *JAMA* 2003;**289**(1):65–9.
25. Miettinen OS. Proportion of disease caused or prevented by a given exposure, trait or intervention. *Am J Epidemiol* 1974;**99**(5):325–32.
26. Atienza AA, King AC. Community-based health intervention trials: an overview of methodological issues. *Epidemiol Rev* 2002;**24**(1):72–9.
27. Koepsell TD, Diehr PH, Cheadle A, et al. Invited commentary: symposium on community intervention trials. *Am J Epidemiol* 1995;**142**(6):594–9.
28. COMMIT Research Group. Community intervention trial for smoking cessation (COMMIT), I: cohort results from a four-year community intervention. *Am J Public Health* 1995;**85**(2):183–92.
29. COMMIT Research Group. Community intervention trial for smoking cessation (COMMIT), II: changes in adult cigarette smoking prevalence. *Am J Public Health* 1995;**85**(2):193–200.

CHAPTER

Case-Control Studies - Part I

12

Much of the effort going into a cohort study involves following up subjects to identify incident cases. Many diseases are uncommon, which means that in order to accrue sufficient cases, sample sizes must be large and the observation period long. This makes for an expensive undertaking, and possibly one that will not deliver results in a timely fashion. In many areas of health care, in particular public health, answers are needed sooner rather than later so that urgent policy decisions can to be made. This argues for an alternative to a cohort-based approach. A response to this need is provided by the case-control design.

When first introduced in the 1950s, the case-control design was initially viewed with skepticism. With subjects recruited on the basis of disease status, and exposure status determined retrospectively, case-control studies were often characterized as a type of "backward" cohort study. This unflattering attitude has persisted to some extent but is no longer warranted. Our understanding of the case-control design has evolved to the point where a well-conducted case-control study should be regarded as the scientific equivalent of a cohort study. In fact, the contemporary view of a case-control study is that it is fundamentally a method of sampling subjects from an underlying cohort, and for that reason shares many of the features of the cohort approach. When sampling is conducted in this fashion, we say that the case-control study is *embedded* in the cohort study.

Suppose a closed cohort study of 1,000 individuals has been conducted and 70 incident cases have occurred. Imagine that as part of the study a blood sample was collected from each subject at the time of enrollment and placed in storage for possible later use. Now suppose it has recently come to the attention of the investigators that a certain exposure which can be measured in the blood might be a risk factor for the disease. The investigators are interested in exploring this connection but the blood test is very expensive. If all 1,000 subjects could be assessed for the exposure, the data from the resulting closed cohort study, which we refer to as closed cohort study #6, would be as in Table 12.1. The investigators have knowledge of only the bottom row of the table; unknown to them, the odds ratio for the cohort study is 3.0.

An efficient alternative to collecting exposure data on the entire cohort is to conduct a case-control study in which "cases" are all incident cases from the cohort study, and "controls" are a probability sample of noncases from the cohort study, say, a 10% simple random sample. By definition, the case-control study is embedded in the cohort study. Analyzing the blood samples for just these individuals, it follows

Epidemiologic Methods. https://doi.org/10.1016/B978-0-44-318780-3.00018-X
Copyright © 2023 Elsevier Inc. All rights reserved.

Table 12.1 Data for closed cohort study #6.

Exposure	Cases	Noncases	Total
$E=1$	60	620	680
$E=0$	10	310	320
Overall	70	930	1,000

Table 12.2 Data for case-control study embedded in closed cohort study #6.

Exposure	Cases	Controls
$E=1$	60	62
$E=0$	10	31
Overall	70	93

from Table 12.1 that the data for the case-control study would (on average) be as in Table 12.2.

Let us focus on the odds ratio as a measure of association in Table 12.2. The odds of a case having had the exposure is $60/10=6.0$, and the odds of a control having had the exposure is $62/31=2.0$. The corresponding odds ratio is $(60/10)/(62/31)=3.0$ and has the following interpretation: developing the disease increases the odds of having had the exposure by a factor of 3.0. This suggests that there is an association between the exposure and the disease, but what the investigators really want to know is whether having had the exposure increases the odds of developing the disease. To that end, let us proceed as if the data in Table 12.2 had come from a closed cohort study. This gives an odds ratio of $(60/62)/(10/31)=3.0$, the same as the previous odds ratio. More than that, both of these odds ratios are the same as the odds ratio computed from the (unknown) cohort data in Table 12.1. It can be shown using Tables 12.1 and 12.2 that this remarkable feature of the odds ratio is not shared by either the risk ratio or risk difference.

The preceding example illustrates a more general phenomenon: data from a case-control study embedded in a closed cohort study can be analyzed with the odds ratio as if the data had come from a closed cohort study. More importantly—and this is the magic of the case-control design—the odds ratio from such an analysis can be given a cohort interpretation. Elucidating this crucial second point is one of the primary goals of this chapter.

The case-control notation used here is basically the familiar closed cohort notation but with a tilde ($\sim$) added overhead. Table 12.3 is the case-control counterpart of Table 4.1.

The odds ratio for a case-control study is defined to be

$$\widetilde{OR}=\frac{\tilde{a}_1/\tilde{b}_1}{\tilde{a}_0/\tilde{b}_0}, \tag{12.1}$$

Table 12.3 Data for case-control study.

Exposure	Cases	Controls
$E=1$	$\tilde{a}_1$	$\tilde{b}_1$
$E=0$	$\tilde{a}_0$	$\tilde{b}_0$
Overall	$\tilde{a}$	$\tilde{b}$

where we note that the "odds" are computed "across the rows", as if the data had come from a closed cohort study.

In all the case-control designs discussed here, the cases are incident, as in the above example. There are case-control designs where the cases are prevalent, but we will not explore this alternative. An essential requirement for all case-control studies is that the selection of cases and controls must be independent of exposure status. If this principle is not followed, selection bias results.

12.1 Case-Control Studies Embedded in Cohort Studies

For a case-control study embedded in a cohort study, case and control sampling can be viewed as a means of efficiently selecting subjects from the underlying cohort, which is then the source population for the case-control study. Usually the cohort study in which the case-control study is embedded has been conducted, but this is not necessary. It is sufficient that the cohort study could be conducted, at least in theory. For each of the case-control designs to be considered, we assume that all incident cases arising in the cohort and a sample of m controls are included in the case-control study. The number m is largely at the discretion of the investigators and could, for instance, equal the number of cases or be some multiple of the number of cases. Thus, the cases are a nonprobability sample, whereas the controls are either a probability sample or a nonprobability sample depending on the design of the case-control study. A more general approach to case selection would be to choose a probability sample of incident cases. In practice, incident cases tend to be in short supply, and so the present assumption is realistic.

Cumulative Case-Control Studies

A *cumulative case-control study* is a type of case-control study that is embedded in a closed cohort study. The controls are a simple random sample of the noncases in the cohort at the end of follow-up. With Table 4.1 representing the data for the (actual or conceptual) closed cohort study, Table 12.4 gives the data for the cumulative case-control study. Because sampling of the m controls is independent of exposure status, $(b_1/b)m$ of them are exposed and $(b_0/b)m$ are unexposed. An alternative way of presenting control data in Table 12.4 is to define the *sampling fraction* for controls to be $f = m/b$, the proportion of noncases selected to be in the control sample. Then $\widetilde{b}_1 = fb_1$, $\widetilde{b}_0 = fb_0$ and $\widetilde{b} = fb$.

Table 12.4 Data for cumulative case-control study.

Exposure	Cases	Controls
$E = 1$	$\widetilde{a}_1 = a_1$	$\widetilde{b}_1 = (b_1/b)m$ $[= fb_1]$
$E = 0$	$\widetilde{a}_0 = a_0$	$\widetilde{b}_0 = (b_0/b)m$ $[= fb_0]$
Overall	$\widetilde{a} = a$	$\widetilde{b} = m$ $[= fb]$

It is easily shown that

$$\widetilde{OR} = OR,$$

where OR is the odds ratio in the underlying cohort. If the risk of becoming a case is "small", we have the approximation $OR \approx RR$, hence

$$\widetilde{OR} \approx RR.$$

Using only data from a cumulative case-control study, it is not possible to compute the risk ratio or risk difference in the underlying cohort. The case-control study in the introduction to this chapter has a cumulative design.

The cumulative case-control design is historically the first type of case-control design that was described. It is perhaps because of the above approximation that there is a common misunderstanding that case-control studies in general require the disease to be "rare". As the above discussion shows, this assumption is not needed for a cumulative case-control study unless an estimate of the risk ratio is desired.[1]

Example: Food Poisoning Study

During February 23–24, 1983 a Caribbean cruise ship experienced an outbreak of gastrointestinal illness characterized by severe vomiting, often accompanied by diarrhea, with symptoms lasting only a few hours.[2] An infectious etiology was suspected. A cumulative case-control study was conducted by administering a standardized questionnaire to all passengers presenting to the on-board physician, as well as a sample of passengers without symptoms. The questionnaire collected information on specific food items eaten at dinner on February 22 and lunch on February 24. There were 53 cases and 48 controls for the dinner component of the study. Treating each food item as a separate exposure, the largest odds ratio by far was for cream pastries ($\widetilde{OR} = 20.6$). No specimens of the cream pastries were available for laboratory evaluation, but swabs of forearm lesions from two of the pastry crew were positive for an enterotoxigenic strain of *Staphylococcus aureus*. Corresponding to the cumulative case-control study is an underlying closed cohort study that was never actually conducted, in which all passengers on the cruise ship were followed for the onset of gastrointestinal illness. ◊

Case-Cohort Studies

A *case-cohort study* is another type of case-control study that is embedded in a closed cohort study. The m controls are a simple random sample of the cohort at the start of follow-up. With this approach to sampling, some individuals selected as controls might later become cases, which makes for confusing terminology. Since the group of controls is a cohort in its own right, let us refer to it as the *subcohort* of the case-cohort study. With Table 4.1 representing the data for the (actual or conceptual) closed cohort study, Table 12.5 gives the data for the case-cohort study. Because sampling of the m controls is independent of exposure status, $(n_1/n)m$ of them are exposed and $(n_0/n)m$ are unexposed. An alternative way of presenting control data in Table 12.5 is to define the sampling fraction for the subcohort to be $f = m/n$, the proportion of the cohort selected to be in the subcohort. Then $\widetilde{b}_1 = fn_1$, $\widetilde{b}_0 = fn_0$ and $\widetilde{b} = fn$.

Table 12.5 Data for case-cohort study.

Exposure	Cases	Subcohort
$E = 1$	$\widetilde{a}_1 = a_1$	$\widetilde{b}_1 = (n_1/n)m\ [= fn_1]$
$E = 0$	$\widetilde{a}_0 = a_0$	$\widetilde{b}_0 = (n_0/n)m\ [= fn_0]$
Overall	$\widetilde{a} = a$	$\widetilde{b} = m\ [= fn]$

It is easily shown that

$$\widetilde{OR} = RR, \tag{12.2}$$

where RR is the risk ratio in the underlying cohort.

Density Case-Control Studies

A *density case-control study* is a type of case-control study that is embedded in an open cohort study. The m controls are a probability sample of the entire cohort at the end of follow-up, where each cohort member has a selection probability proportional to the person-time experienced by that individual. For example, someone with five years of follow-up would have a selection probability that is five times the selection probability of someone with only one year of follow-up. As in the case-cohort design, a control might also be a case. With Table 4.4 representing the data for the (actual or conceptual) open cohort study, Table 12.6 gives the data for the density case-control study. Because sampling of the m controls is independent of exposure status, $(y_1/y)m$ of them are exposed and $(y_0/y)m$ are unexposed.

It is easily shown that

$$\widetilde{OR} = IR, \tag{12.3}$$

where IR is the incidence rate ratio in the underlying cohort.

Table 12.6 Data for density case-control study.

Exposure	Cases	Controls
$E=1$	$\tilde{a}_1 = a_1$	$\tilde{b}_1 = (y_1/y)m$
$E=0$	$\tilde{a}_0 = a_0$	$\tilde{b}_0 = (y_0/y)m$
Overall	$\tilde{a} = a$	$\tilde{b} = m$

Nested Case-Control Studies

A *nested case-control study* is another type of case-control study that is embedded in an open cohort study.[3] Unlike the density design, with this approach it is not necessary to wait until the end of follow-up to select controls. As each case arises, one or more controls are selected from cohort members who are noncases at that time. In this way, the case and corresponding controls, which together form what is called a *risk set*, are "matched" on follow-up time and possibly other variables. The risk set approach to selecting cases and controls has a number of characteristic features: a subject can be selected as a control for more than one risk set; a subject who is a control in one risk set and subsequently develops the disease becomes the case in a corresponding risk set; a subject who becomes a case is no longer eligible to be a control. The statistical analysis of data from a nested case-control study typically relies on regression methods that yield an estimate of the hazard ratio.

Example: Physicians' Health Study

The Physicians' Health Study was a double-blind, placebo-controlled, 2×2 factorial randomized clinical trial conducted to investigate whether low-dose aspirin decreases cardiovascular mortality and whether beta carotene reduces the incidence of cancer.[4,5] A total of 22,0171 U.S. male physicians 40–84 years old with no history of heart attack, stroke or cancer and who met additional inclusion criteria were randomized to receive one of four treatments: active aspirin and active beta carotene; active aspirin and beta carotene placebo; aspirin placebo and active beta carotene; aspirin placebo and beta carotene placebo. There were 11,037 subjects in the active aspirin group and 11,034 in the aspirin placebo group. Based on an intention to treat analysis, the aspirin component of the trial was terminated early after an average follow-up of about five years due to a beneficial effect of aspirin on the risk of fatal and nonfatal heart attack.

Blood samples collected at the start of follow-up from all subjects in the Physicians' Health Study were used to conduct a nested case-control study that investigated whether a mutation of the gene coding for a certain coagulation protein increases the risk of heart attack, stroke and deep venous thrombosis/pulmonary embolism.[6] Cases were subjects with a confirmed diagnosis of any of the preceding three conditions, of which there were 374, 209 and 121, respectively. Each case was pair-matched to a control using risk-set sampling, with matching on time since randomization, age and smoking habits. The key finding was that the gene mutation was associated with an increased risk of deep venous thrombosis/pulmonary embolism but not heart attack or stroke. ◊

Case-control studies embedded in cohort studies have a place in epidemiology, as the food poisoning study and Physicians' Health Study illustrate, but in practice, the majority of case-control studies are conducted using quite different methods.[3,7] In the next two sections, we consider the two most common types of case-control studies—population-based case-control studies and hospital-based case-control studies. Much of the discussion relates to issues around the appropriate selection of controls.[8–12, 13(p.79–82), 14] For brevity, we usually abbreviate the terms population-based and hospital-based to P-B and H-B, respectively.

12.2 Population-Based Case-Control Studies

In a P-B case-control study, the source population is either a geographically-defined population or some entity that behaves in a similar fashion, such as the workforce of a large multinational business.[3] If we view the source population as an open cohort with staggered entry that is observed over a given time period, a P-B case-control study can be thought of as a nested case-control study embedded in an open cohort study. By definition, the source population for a P-B case-control study (at a given time) is comprised of individuals at risk of becoming a case of the disease (at that time).

Selection of Cases

The most convenient mechanism for identifying cases for a P-B case-control study is a disease registry. The source population is then all individuals residing in the catchment area of the registry at risk of the disease. The cases are usually a consecutive series or some other nonprobability sample of individuals recently placed on the registry. Most disease registries are designed to collect information on cancers, notifiable infections or congenital anomalies. Someone newly-diagnosed with cancer is usually regarded as being an incident case. On the other hand, someone newly-diagnosed with a congenital anomaly, even one diagnosed at birth, is inherently a prevalent case, as discussed in Section 4.1 in connection with the Down syndrome study.

Selection of Controls

According to the nested case-control approach, as each case is identified, a sample of controls needs to be selected for the risk set corresponding to that case. If a roster of individuals residing in the catchment area of the disease registry is available, such as a recent census, controls can be selected using probability sampling with or without matching on (calendar) time and possibly other variables. When no roster is available, it is necessary to turn to other methods.

In *random-digit dialing*, lists of residential area codes for the source population are obtained from phone companies.[8,11] These are augmented with random sequences of digits to form phone numbers that are used to contact individuals or households. Once contact is established, an attempt is made to recruit one or more eligible controls into the study. As can be imagined, there are numerous obstacles to the success of such an undertaking. For example, some households might not have phone service, call screening could limit access to potential respondents, and the mobility of the general population may make it unclear which area codes should be included. All of these problems, and others, raise concerns about the representativeness of the control sample.

In *neighborhood sampling*, when a case is recruited, the "neighborhood" where that person lives is identified and an attempt is made to recruit one or more individuals residing there.[8,11] For example, households within a fixed distance of the index household could be considered eligible and a selection grid used to identify potential controls living at a given address. Neighborhood sampling tends to match cases and controls on socioeconomic and other factors that are characteristic of the area in which they live.

12.3 Hospital-Based Case-Control Studies

Selection of Cases

When there is no registry available for the disease of interest, an alternative method of identifying cases must be found. This usually involves some type of health care network, such as a group of outpatient medical clinics or acute care hospitals. For simplicity, we focus on a single hospital. In an H-B case-control study, the cases are usually a nonprobability sample of hospital patients newly-diagnosed with a given disease.[3] In order for cases to be regarded as incident (in some source population), we consider only diseases that require prompt hospitalization once they are suspected or diagnosed. (For conditions that are generally not serious enough to require hospitalization, we can replace "hospital" with "medical clinic" in what follows.)

It may be that the hospital has a well-defined catchment area, allowing it to be viewed as a disease registry. In this situation, the source population for cases is comprised of all individual residing in the catchment area of the hospital who are at risk of the disease and have access to the hospital. Usually circumstances are not so convenient and we are left to imagine who would be treated in the hospital and under what conditions. For instance, if the hospital is a cancer treatment center that attracts patients from around the world, what exactly is the population it serves, and how are we to know who would be treated there if treatment were needed? Nevertheless, we can imagine that such a group of individuals exists and that it has features similar to a geographically-defined population. In this way, the source population for cases is defined secondarily to the cases enrolled in the case-control study.[9,10, 13(p.54–55,79)]

Selection of Controls

If the hospital has the characteristics of a disease registry, we can select controls using methods developed for P-B case-control studies, such as probability sampling from a roster, random-digit dialing and neighborhood sampling. When the hospital does not fit the disease registry paradigm, there is no guarantee that these methods will identify individuals who would be treated in the hospital if treatment were needed. The usual method of recruiting controls for an H-B case-control study automatically ensures that the preceding condition is satisfied. Specifically, controls are a nonprobability sample of hospital patients newly-diagnosed with some disease other than the case disease.[8,11] Similar to cases, in order for the controls to be regarded as incident (in some source population), only diseases that require prompt hospitalization are considered. Also similar to cases, the source population for controls is defined secondarily to the controls enrolled in the case-control study.[9,10, 13(p.79–82), 14]

In order to meaningfully compare cases and controls, we need the source population for cases and the source population for controls to be the same.[9, 13(p.79), 14] In other words, we require that individuals in the source population for cases are all those who would be admitted to the hospital for the control disease were it to develop, and similarly, that individuals in the source population for controls are all those who would be admitted to the hospital for the case disease were it to develop. When expressed this way, assuming that there is a common source population seems reasonable and, aside from occasional discrepancies, likely to be satisfied in practice. As an exception, there may be individuals in the source population for cases who are not at risk of the control disease, and conversely. The study design can be structured to increase the likelihood that the case and control source populations are the same. For example, cases and controls could be restricted to individuals residing in a geographic area surrounding the hospital. In any event, we proceed on the basis that there is a common source population for cases and controls, and refer to it simply as the *source population* for the case-control study.

The above discussion is framed in terms of a single control group, but more than one control group can be used, each defined in terms of a distinct disease. If the odds ratios estimated using different control groups are similar, the findings from the study are strengthened. On the other hand, when the odds ratio estimates are dissimilar, the investigator has the difficult task of explaining such a finding.[13(p.82)] As will be made clear below, it is important that the disease(s) used to define the control group(s) are restricted to ones that do not have the exposure as a risk factor.[9,11] For example, smoking causes coronary heart disease and chronic obstructive pulmonary disease. These diseases should not be used to define control groups in an H-B case-control study of smoking and lung cancer.

12.4 Case-Control Studies in a Stationary Population

We now assume that the source population for a case-control study is comprised of an exposed population and an unexposed population, both of which are stationary.

This means that the source population itself is stationary.[3] In what follows, we adopt the notation of Chapter 5.

Population-Based Case-Control Studies

In a P-B case-control study, A incident cases occur during the observation period and M individuals are at risk at any time during the observation period. Because of the stationary property, controls can be sampled in a cross-sectional manner. We assume that a simple random sample of m controls is selected. If sampling is performed prior to the end of the observation period, there is the possibility that some controls will later become cases. In a large population the chances of this happening are small. Nevertheless, this situation can be avoided by taking full advantage of the stationary property and selecting the control sample at the end of the observation period. Table 12.7 gives the data for a P-B case-control study. Because sampling of the m controls is independent of exposure status, $(M_1/M)m$ of them are exposed and $(M_0/M)m$ are unexposed. An alternative way of presenting control data in Table 12.7 is to define the sampling fraction for controls to be $f = m/M$, the proportion of the at-risk population selected to be in the control sample. Then $\widetilde{b}_1 = fM_1$, $\widetilde{b}_0 = fM_0$ and $\widetilde{b} = fM$.

Table 12.7 Data for population-based case-control study.

Exposure	Cases	Controls
$E=1$	$\widetilde{a}_1 = A_1$	$\widetilde{b}_1 = (M_1/M)m\ [= fM_1]$
$E=0$	$\widetilde{a}_0 = A_0$	$\widetilde{b}_0 = (M_0/M)m\ [= fM_0]$
Overall	$\widetilde{a} = A$	$\widetilde{b} = m\ [= fM]$

It is easily shown that[15]

$$\widetilde{OR} = IR, \tag{12.4}$$

where IR is the incidence rate ratio in the stationary population.

Hospital-Based Case-Control Studies

For an H-B case-control study, there are A hospital patients newly-diagnosed during the observation period with the case disease. Let us add a dagger ($\dagger$) to the notation to indicate the control disease. Accordingly, there are $A^{\dagger}$ hospital patients newly-diagnosed during the observation period with the control disease. Table 12.8 gives the data for an H-B case-control study.

It is shown in Appendix J that

$$\widetilde{OR} = \frac{IR}{IR^{\dagger}}, \tag{12.5}$$

where IR is the incidence rate ratio for the exposure-case disease association in the stationary population, and $IR^{\dagger}$ is the corresponding incidence rate ratio for the exposure-control disease association. The derivation of (12.5) depends crucially on the assumption that there is a common source population for cases and controls.

The measure of effect of interest in an H-B case-control study is IR. We see from (12.5) that $\widetilde{OR} = IR$ if and only if $IR^{\dagger} = 1$, that is, if and only if the exposure is not a risk factor for the control disease. This provides a justification for the remark made earlier that the disease(s) used to define the control group(s) should be restricted to ones that do not have the exposure as a risk factor.

Table 12.8 Data for hospital-based case-control study.

Exposure	Cases	Controls
$E = 1$	$\widetilde{a}_1 = A_1$	$\widetilde{b}_1 = A_1^{\dagger}$
$E = 0$	$\widetilde{a}_0 = A_0$	$\widetilde{b}_0 = A_0^{\dagger}$
Overall	$\widetilde{a} = A$	$\widetilde{b} = A^{\dagger}$

Example: Heart Attack Study

An H-B case-control was conducted during 1976–1978 to investigate the association between recent oral contraceptive use and heart attack.[16] Cases were 234 premenopausal women 25–49 years old admitted to one of 155 hospitals in Massachusetts, Pennsylvania and New York because of a first heart attack. Controls were 1,742 patients with no history of heart attack selected from the surgical, orthopedic and medical services of the hospitals from which cases were recruited, and with a primary diagnosis judged to be unrelated to oral contraceptive use or smoking. The crude odds ratio of $\widetilde{OR} = 1.68$ suggests that recent oral contraceptive use increases the incidence rate of first heart attack in premenopausal women. ◊

12.5 Ratio of Controls to Cases

All other things being equal, the larger the sample size of a case-control study, the smaller the variance of the odds ratio estimate. The availability of cases is often limited, whereas potential controls are frequently plentiful. A common practice is to recruit as many cases as possible along with an equal or greater number of controls. The *ratio of controls to cases* is defined to be the number of controls in the study relative to the number of cases. For a given number of cases, as the ratio of controls to cases increases, the variance of the odds ratio estimate decreases. But there is a diminishing rate of return, and little is gained by increasing the ratio beyond 4:1.[12,17]

References

1. Greenland S, Thomas DC. On the need for the rare disease assumption in case-control studies. *Am J Epidemiol* 1982;**116**(3):547–53.
2. Waterman SH, Demarcus TA, Wells JG, et al. Staphylococcal food poisoning on a cruise ship. *Epidemiol Infect* 1987;**99**(2):349–53.
3. Vandenbroucke JP, Pearce N. Case-control studies: basic concepts. *Int J Epidemiol* 2012;**41**(5):1480–9.
4. Steering Committee of the Physicians' Health Study Research Group. Preliminary report: findings from the aspirin component of the ongoing Physicians' Health Study. *N Engl J Med* 1988;**318**(4):262–4.
5. Steering Committee of the Physicians' Health Study Research Group. Final report on the aspirin component of the ongoing Physicians' Health Study. *N Engl J Med* 1989;**321**(3):129–35.
6. Ridker PM, Hennekens CH, Lindpaintner K, et al. Mutation in the gene coding for coagulation factor V and the risk of myocardial infarction, stroke, and venous thrombosis in apparently healthy men. *N Engl J Med* 1995;**332**(14):912–7.
7. Knol MJ, Vandenbroucke JP, Scott P, et al. What do case-control studies estimate? Survey of methods and assumptions in published case-control research. *Am J Epidemiol* 2008;**168**(9):1073–81.
8. Grimes DA, Schultz KF. Compared to what? Finding controls for case-control studies. *Lancet* 2005 Apr 16;**365**(9468):1429–33.
9. Miettinen OS. The "case-control study": valid selection of subjects. *J Chron Dis* 1985;**38**(7):543–8.
10. Wacholder S, McLaughlin JK, Silverman DT, et al. Selection of controls in case-control studies, I: principles. *Am J Epidemiol* 1992;**135**(9):1019–28.
11. Wacholder S, Silverman DT, McLaughlin JK, et al. Selection of controls in case-control studies, II: types of controls. *Am J Epidemiol* 1992;**135**(9):1029–41.
12. Wacholder S, Silverman DT, McLaughlin JK, et al. Selection of controls in case-control studies, III: design options. *Am J Epidemiol* 1992;**135**(9):1042–50.
13. Miettinen OS. *Theoretical epidemiology: principles of occurrence research in medicine.* New York: John Wiley & Sons; 1985.
14. Miettinen OS. Response: the concept of secondary base. *J Clin Epidemiol* 1990;**43**(9):1017–20.
15. Miettinen O. Estimability and estimation in case-referent studies. *Am J Epidemiol* 1976;**103**(2):226–35.
16. Shapiro S, Slone D, Rosenberg L, et al. Oral-contraceptive use in relation to myocardial infarction. *Lancet* 1979 Apr 7;**1**(8119):743–6.
17. Ury HK. Efficiency of case-control studies with multiple controls per case: continuous or dichotomous data. *Biometrics* 1975;**31**(3):643–9.

CHAPTER

Case-Control Studies - Part II

13

13.1 Bias in Case-Control Studies

Selection Bias

Berkson's bias, named after Joseph Berkson (1899–1982), an American physician and statistician, is a family of selection biases that occur in H-B case-control studies.[1–8] In one of its several manifestations, exposure is a cause of the control disease but not the case disease in the general population, and yet exposure is associated with the case disease in hospitalized patients. From the causal diagram in Fig. 13.1, we see that this type of Berkson's bias can be interpreted as a form of collider bias.[5,6,8] In his original analysis, Berkson was concerned with prevalent disease. When cases and controls are restricted to patients with incident disease, Berkson's bias is typically small and usually of limited practical importance.[4–6]

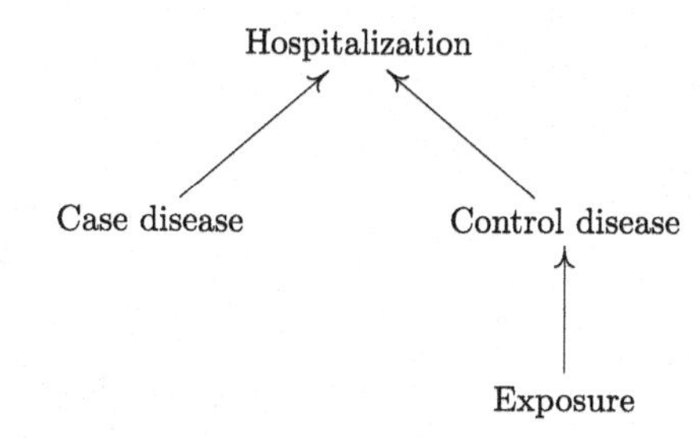

FIGURE 13.1

Causal diagram for Berkson's bias.

Information Bias

Recall bias is a type of information bias that occurs in case-control studies when the accuracy with which exposure history is recalled differs for cases and controls. For example, someone who has recently been diagnosed with a disease might ruminate over what could have been its cause and consequently come to have a detailed recollection of past exposures, whereas someone without the disease will not be preoccupied in the same way. Recall bias results in differential misclassification of exposure status.

Epidemiologic Methods. https://doi.org/10.1016/B978-0-44-318780-3.00019-1
Copyright © 2023 Elsevier Inc. All rights reserved.

Nondifferential Misclassification of Exposure Status

Consider a cumulative case-control study in which the exposure is dichotomous and subject to nondifferential misclassification. By definition, the "test" used to determine exposure status has the same (unknown) sensitivity (S) and specificity (T) in cases and controls. Let us investigate the impact of this misclassification on the odds ratio (OR). By definition, the case-control study is embedded in a closed cohort study. We assume that $0 < S < 1$ and $0 < T < 1$ and $S + T > 1$, and also that $0 < R_0 < 1$ and $0 < R_1 < 1$. These conditions are almost always satisfied in practice.

Keeping with the star notation of Section 3.2, we denote the misclassified odds ratio by $\widetilde{OR}^\star$. It is shown in Appendix K that $\widetilde{OR}^\star = 1$ if and only if $OR = 1$. It is also shown in Appendix K that if $OR \neq 1$, then

$$OR > \widetilde{OR}^\star > 1 \quad \text{or} \quad OR < \widetilde{OR}^\star < 1. \tag{13.1}$$

Thus, when the exposure is dichotomous, nondifferential misclassification of exposure status in a cumulative case-control study biases the odds ratio estimate toward the null value. However, when the exposure has more than two levels, the odds ratio estimate can be biased away from the null value, even when misclassification is nondifferential.[9,10]

Example: Nondifferential Misclassification of Exposure Status in a Cumulative Case-Control Study

Table 13.3(a) gives the data for closed cohort study #7. The risk in the unexposed cohort is $R_0 = 0.20$ and the crude odds ratio is $OR = 2.67$. Table 13.3(b) gives the data for a cumulative case-control study embedded in closed cohort study #7, where the sampling fraction for controls is 50% and where there is no misclassification of exposure status. The odds ratio is $\widetilde{OR} = 2.67$, as expected.

Now suppose that when the cumulative case-control study is conducted, there is nondifferential misclassification of exposure status, with a sensitivity of 80% and specificity of 90%. Tables 13.3(c) and 13.3(d) give the misclassified data for cases and controls, respectively, where computations are performed as in Table 3.2. The far-right columns of Tables 13.3(c) and 13.3(d) become the columns of Table 13.3(e), which gives the misclassified data for the case-control study. The misclassified odds ratio is $\widetilde{OR}^\star = 1.96$, which is biased toward the null value, as expected. ◊

Consider a cumulative case-control study embedded in a closed cohort study where the exposed and unexposed cohorts have the same number of subjects, that is, $n_1 = n_0$, and where there is nondifferential misclassification of exposure status. Fig. 13.2 shows the ratio $\widetilde{OR}^\star/OR$ as a function of R_0 for $OR = 2, 5, 10$, where the sensitivity and specificity are fixed at 80% and 90%, respectively. The formula for $\widetilde{OR}^\star/OR$ used in Fig. 13.2 is given in Appendix K. Since each of the selected values of OR is greater than 1, it follows from (13.1) that $\widetilde{OR}^\star/OR < 1$ for all values of R_0, and this is reflected in the graph. For a given value of R_0, the bias due to nondifferential misclassification increases as OR increases.

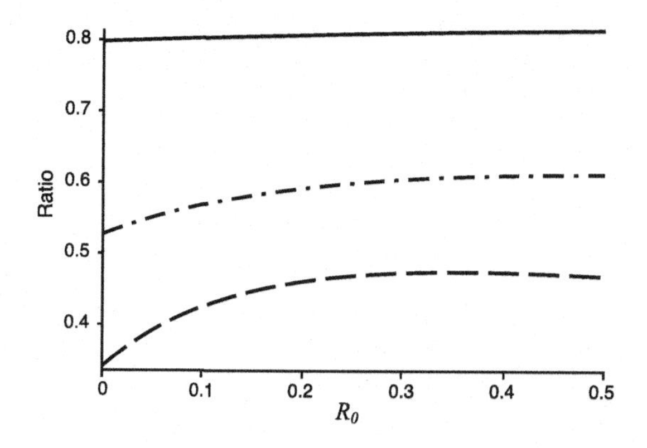

FIGURE 13.2

$\widetilde{OR}^{\star}/OR$ for a cumulative case-control study with nondifferential misclassification of exposure status as a function of R_0: sensitivity 80% and specificity 90% ($OR = 2$: *solid*; $OR = 5$: *dashdot*; $OR = 10$: *dash*).

13.2 Stratification in Case-Control Studies

Tables 13.1 and 13.2 give the stratum-specific data for a case-control study. They are the case-control counterparts to Tables 7.2 and 7.3 for a closed cohort study.

Table 13.1 Stratum-specific data for case-control study.

Exposure	Cases	Controls
$E=1$	$\widetilde{a}_{1i}$	$\widetilde{b}_{1i}$
$E=0$	$\widetilde{a}_{0i}$	$\widetilde{b}_{0i}$

Table 13.2 Crude and stratum-specific data for case-control study.

Stratum	Cases		Controls	
	$E=1$	$E=0$	$E=1$	$E=0$
$F=k$	$\widetilde{a}_{1k}$	$\widetilde{a}_{0k}$	$\widetilde{b}_{1k}$	$\widetilde{b}_{0k}$
$\vdots$	$\vdots$	$\vdots$	$\vdots$	$\vdots$
$F=i$	$\widetilde{a}_{1i}$	$\widetilde{a}_{0i}$	$\widetilde{b}_{1i}$	$\widetilde{b}_{0i}$
$\vdots$	$\vdots$	$\vdots$	$\vdots$	$\vdots$
$F=0$	$\widetilde{a}_{10}$	$\widetilde{a}_{00}$	$\widetilde{b}_{10}$	$\widetilde{b}_{00}$
Overall	$\widetilde{a}_{1}$	$\widetilde{a}_{0}$	$\widetilde{b}_{1}$	$\widetilde{b}_{0}$

Effect Modification

Effect modification of the odds ratio in a case-control study is defined similarly to effect modification of the risk ratio and risk difference in a closed cohort study. See Section 6.1.

Standardized Odds Ratio

For the remainder of this chapter, we focus on the case-cohort and P-B case-control designs—the former because it is useful for illuminating the behavior of case-control studies in general, and the latter because it is the most frequently used case-control design.[11] We assume in what follows that the source population for the P-B case-control study is comprised of exposed and unexposed populations, both of which are stationary.

Let us begin with a few definitions. For both case-cohort and P-B case-control studies, the *observed number* of exposed cases is $\widetilde{a}_1$, and the *expected number* of exposed cases is defined to be[12]

$$\widetilde{a}_{0(1)} = \sum_{i=0}^{k} \frac{\widetilde{a}_{0i}\widetilde{b}_{1i}}{\widetilde{b}_{0i}} . \tag{13.2}$$

It is shown in Appendix K that for case-cohort studies,

$$\widetilde{a}_{0(1)} = a_{0(1)} , \tag{13.3}$$

and for P-B case-control studies,

$$\widetilde{a}_{0(1)} = A_{0(1)} , \tag{13.4}$$

where $a_{0(1)}$, the expected number of incident cases in the exposed cohort, is defined by (7.5), and $A_{0(1)}$, the expected number of incident cases in the exposed population, is defined by (5.17).

For both case-cohort and P-B case-control studies, the standardized odds ratio is defined to be[12]

$$s\widetilde{OR} = \frac{\widetilde{a}_1}{\widetilde{a}_{0(1)}} . \tag{13.5}$$

It is shown in Appendix K that $s\widetilde{OR}$ is a weighted average of stratum-specific odds ratios. It is also shown in Appendix K that for case-cohort studies,

$$s\widetilde{OR} = sRR , \tag{13.6}$$

and for P-B case-control studies,

$$s\widetilde{OR} = sIR , \tag{13.7}$$

where sRR, the standardized risk ratio in the closed cohort, is defined by (7.4), and sIR, the indirectly standardized incidence rate ratio in the population, is defined by (5.15).

Mantel-Haenszel Odds Ratio

We would be remiss were we not to mention the venerable *Mantel-Haenszel odds ratio* for case-control studies, defined by[13]

$$\widetilde{OR}_{\text{MH}} = \sum_{i=0}^{k} \frac{\tilde{a}_{1i}\tilde{b}_{0i}}{\tilde{n}_i} \bigg/ \sum_{i=0}^{k} \frac{\tilde{a}_{0i}\tilde{b}_{1i}}{\tilde{n}_i},$$

where $\tilde{n}_i$ is the number of distinct subjects in stratum i. This measure of effect ushered in the family of Mantel-Haenszel measures of effect that were foundational to the development of epidemiology, and which are still used to analyze data from a range of case-control designs.[14(p.415–445), 15(p.187–188)]

13.3 Confounding in Case-Control Studies

Confounding in Case-Cohort Studies

We say that confounding is present in a case-cohort study precisely when there is confounding in the closed cohort study in which the case-cohort study is embedded. The question that arises is whether data from a case-cohort study can be used to determine if there is confounding in the underlying closed cohort study.

Counterfactual Confounding

It is shown in Appendix K that $R_0 \neq R_{0(1)}$ is equivalent to

$$\text{(C21)} \quad \widetilde{OR} \neq s\widetilde{OR}.$$

So we can decide if (C7) is satisfied using case-cohort data.

Classical Confounding

We have from Table 12.5 and its stratum-specific counterpart that

$$\frac{\tilde{b}_{1i}}{\tilde{b}_1} = \frac{n_{1i}}{n_1}$$

$$\frac{\tilde{b}_{0i}}{\tilde{b}_0} = \frac{n_{0i}}{n_0}.$$

Thus, it is possible to determine if (C13) is satisfied based on case-cohort data.

It is shown in Appendix K that

$$\frac{\widetilde{a}_{0i}/\widetilde{b}_{0i}}{\widetilde{a}_{0j}/\widetilde{b}_{0j}} = \frac{R_{0i}}{R_{0j}}, \tag{13.8}$$

where we note that the left-hand side of (13.8) is the odds ratio in unexposed subjects comparing stratum $F = i$ to stratum $F = j$. Setting $j = 0$, it follows from (13.8) that $R_{00}, \ldots, R_{0i}, \ldots, R_{0k}$ are not all equal if and only if

$$\frac{\widetilde{a}_{0i}/\widetilde{b}_{0i}}{\widetilde{a}_{00}/\widetilde{b}_{00}} \neq 1$$

for at least one value of i. We can therefore decide whether (C14) is satisfied using case-cohort data.

Example: Case-Cohort Study Embedded in Closed Cohort Study #4

Table 13.4(a) gives the data for a case-cohort study that is embedded in closed cohort study #4, where the sampling fraction for the subcohort is 10%. The entries in Table 13.4(a) are obtained from Table 9.2(a) by selecting all cases from the cohort study to get the cases for the case-cohort study, and by selecting a 10% simple random sample of the cohort at the start of follow-up to get the subcohort. See the computations below Table 13.4(a). Applying the above methods to the case-cohort study, we find from Tables 13.4(b) and 13.4(c) that there is counterfactual and classical confounding (of the odds ratio). This is consistent with the findings presented in Section 9.3 for closed cohort study #4, where we found counterfactual and classical confounding (of the risk ratio). As an illustration of the computations involved, we have $\widetilde{a}_1 = 1,800$ and

$$\widetilde{a}_{0(1)} = \left(\frac{80 \times 200}{800}\right) + \left(\frac{40 \times 800}{200}\right) = 180,$$

hence

$$s\widetilde{OR} = \frac{1,800}{180} = 10.0.$$

Referring to Table 9.2(c), we note that (13.6) is satisfied. ◊

Confounding in Population-Based Case-Control Studies

Similar to a case-cohort study, we say there is confounding in a P-B case-control study precisely when there is confounding in the underlying stationary population. As before, the issue is whether data from a P-B case-control study can be used to determine if there is confounding in the underlying population.

Classical Confounding

We have from Table 12.7 and its stratum-specific counterpart that

$$\frac{\widetilde{b}_{1i}}{\widetilde{b}_1} = \frac{M_{1i}}{M_1}$$

$$\frac{\widetilde{b}_{0i}}{\widetilde{b}_0} = \frac{M_{0i}}{M_0}.$$

Thus, it is possible to determine if (C19) is satisfied based on P-B case-control data.

It is shown in Appendix K that

$$\frac{\widetilde{a}_{0i}/\widetilde{b}_{0i}}{\widetilde{a}_{0j}/\widetilde{b}_{0j}} = \frac{I_{0i}}{I_{0j}}, \quad (13.9)$$

from which it follows that $I_{00}, \ldots, I_{0i}, \ldots, I_{0k}$ are not all equal if and only if

$$\frac{\widetilde{a}_{0i}/\widetilde{b}_{0i}}{\widetilde{a}_{00}/\widetilde{b}_{00}} \neq 1$$

for at least one value of i. We can therefore decide whether (C20) is satisfied using P-B case-control data.

13.4 Controlling Confounding in Case-Control Studies

Similar to cohort studies, methods to control confounding in a case-control study can be implemented at the design stage or when the data are being analyzed. Randomization plays no role in case-control studies because exposure status has already determined by the time subjects are recruited. Restriction in case-control studies is similar to restriction in cohort studies. Standardization[14,16] and stratification (e.g. Mantel-Haenszel methods)[14] were discussed in Section 13.2. Regression (e.g. logistic regression)[17,18] and inverse probability weighting[19] are beyond the scope of this book. That leaves matching, which we now discuss.

Matching in Case-Control Studies

Matching in a cohort study involves matching exposed and unexposed subjects, whereas matching in a case-control study involves matching cases and controls. More specifically, in a matched case-control study, controls are selected so as to have the same distribution of matching variables as the cases. As a result, controls in a matched case-control study are not a simple random sample of all possible controls. Rather, they can be thought of as a stratified random sample of all possible controls, where the selection of controls in each stratum is determined by the distribution of matching variables in the cases. In Sections 12.1 and 12.4, we assumed that the subcohort for a case-cohort study is a simple random sample of the cohort, and the controls for a P-B case-control study in a stationary population are a simple random sample of the at-risk population. Therefore, as it stands, the discussion of confounding in Section 13.3 does not apply to matched case-cohort and matched P-B case-control studies.

Example: Matched Case-Cohort Study Embedded in Closed Cohort Study #4 (Matching Variable is a Risk Factor for the Disease)

We see from Tables 9.2(a) and 9.2(b) that in closed cohort study #4, F is a risk factor for the disease, and F is associated with E. Table 13.5(a) gives the data for a frequency-matched case-cohort study embedded in closed cohort study #4, where there are equal numbers of cases and members of the subcohort. The entries in Table 13.5(a) are obtained from Table 9.2(a) as follows. First, select all cases from the cohort study to get the cases for the case-cohort study. This gives 1,640 and 280 cases in the strata of F for the case-cohort study. Second, select an equal number of subjects from the strata of F in the cohort at the start of follow-up. This gives 1,640 and 280 subjects in the strata of F for the subcohort. (This is the step that introduces frequency matching). Third, determine the number of exposed and unexposed subjects in the subcohort using the numbers of exposed and unexposed subjects in each stratum of F in the cohort at the start of follow-up. See the computations below Table 13.5(a).

Table 13.5(b) gives the odds ratios based on Table 13.5(a). As an illustration of the computations involved, we have $\widetilde{a}_1 = 1,800$ and

$$\widetilde{a}_{0(1)} = \left(\frac{80 \times 56}{224}\right) + \left(\frac{40 \times 1,312}{328}\right) = 180,$$

hence

$$s\widetilde{OR} = \frac{1,800}{180} = 10.0.$$

From Table 9.2(c), there is counterfactual confounding in cohort study #4. We take the standardized risk ratio ($sRR = 10.0$) as an overall measure of effect for the cohort study. The crude risk ratio ($RR = 15.0$) for the cohort study is biased away from the null value. From Table 13.5(b), we observe that $s\widetilde{OR} = sRR$ and that the crude odds ratio ($\widetilde{OR} = 6.1$) for the case-cohort study is biased toward the null value. ◊

Matching in a case-control study on a variable that is associated with the exposure introduces a selection bias "into the data".[20] To see this, let us denote the exposure, disease and matching variables by E, D and F, respectively. If E and F are positively (negatively) associated in the cases and positively (negatively) associated in those members of the source population at risk of becoming a case, matching on F makes the distribution of E in the controls more like the distribution of E in the cases than it would be in an unmatched case-control study, and this biases the crude odds ratio for the E-D association toward the null value. On the other hand, if E and F are positively (negatively) associated in the cases and negatively (positively) associated in those members of the source population at risk of becoming a case, matching on F makes the distribution of E in the controls less like the distribution of E in the cases than it would be in an unmatched case-control study, and this biases the crude odds ratio for the E-D association away from the null value.

In practice, when E and F are associated, it is usually the first of these scenarios that applies, that is, E and F are positively associated in both cases and those members of the source population at risk of becoming a case, or E and F are negatively associated in both cases and those members of the source population at risk of becoming a case. Either way, the crude odds ratio for the E-D association is biased toward the null value. In the extreme situation where E and F are perfectly correlated, matching on F is effectively matching on E, which results in a crude odds ratio of 1.

As an example of this phenomenon, the Doll and Hill lung cancer study discussed in Section 1.6 is a matched H-B case-control study, with matching on age and other variables. Since older age is associated with both cigarette smoking and lung cancer mortality, it follows that the crude odds ratio ($\widetilde{OR} = 2.7$) for the smoking-lung cancer association is likely an underestimate of the true value.

The selection bias that results from matching in a case-control study behaves like confounding in that it can be controlled by an appropriate statistical analysis. But it can be difficult to distinguish the two phenomena on the basis of study data. We observed in the above example that in the underlying cohort, the crude risk ratio is biased away from the null value, but in the case-cohort study, the crude odds ratio is biased toward the null value. The reason for the change in direction of bias is that the selection bias due to matching in the case-cohort study overwhelms the bias due to confounding in the cohort study.

Matching in a case-control study results in selection bias even when the matching variable is not a risk factor for the disease.[20] We illustrate this with a matched case-control study embedded in a closed cohort study where the matching variable is associated with the exposure but not with the disease.

Example: Matched Case-Cohort Study Embedded in Closed Cohort Study #8 (Matching Variable is not a Risk Factor for the Disease)

Tables 13.6(a)–13.6(c) present closed cohort study #8. We observe from Tables 13.6(a) and 13.6(b) that F is associated with E, and that F is not a risk factor for the disease. There is no counterfactual or classical confounding. Accordingly, the crude risk ratio ($RR = 10.0$) is an overall measure of effect for the cohort study. Table 13.7(a) gives the data for a frequency-matched case-cohort study embedded in closed cohort study #8, where there are equal numbers of cases and members of the subcohort. We see from Table 13.7(b) that the standardized odds ratio ($s\widetilde{OR} = 10.0$) differs from the crude odds ratio ($\widetilde{OR} = 4.8$). The reason is that a selection bias was introduced by matching on F, even though F is not a risk factor for the disease. The selection bias behaves like confounding in that it is eliminated by standardization. So the standardized odds ratio for the case-cohort study equals the crude risk ratio for the cohort study. ◊

Statistical Efficiency of Matching in Case-Control Studies

Based on our experience with matching in cohort studies, it is reasonable to expect that the goal of matching in case-control studies is control of confounding. As we have just seen, matching on a confounder introduces a selection bias into case-control data that can be addressed, for example, by standardization. Thus, when confounding is present, this type of data analysis is called for whether the case-control study is matched or not. It seems that a more straightforward approach to controlling confounding would be to conduct an unmatched case-control study in the first place. However, the reason that a matched design should be considered is that the aim of matching in a case-control study is not to control confounding but rather to improve *statistical efficiency*, as we now discuss.

In statistics, efficiency is measured by comparing variances. For a given sample size and measure of effect, a study design that yields a smaller variance compared to an alternative design is said to be more statistically efficient. Let us compare, using heuristic arguments, the statistical efficiency of unmatched and matched case-control designs. In an unmatched case-control study, the confounder distribution differs for cases and controls, whereas in a matched case-control study, it is the same (or close to the same). As a result, in an unmatched case-control study, some confounder strata may have relatively few cases compared to controls, and others may have relatively few controls compared to cases. Small numbers of subjects in strata translate into a large overall variance. As a consequence, an unmatched case-control design tends to be less statistically efficient than a matched case-control design.

It has been shown that the gain in statistical efficiency from a matched design is directly related to the strength of the association between the confounder and the disease, and inversely related to the strength of the association between the exposure and the disease.[21] More specifically, unless the association between the confounder and disease is strong, the gain in statistical efficiency is likely to be small.[22] Even when there are potential gains in statistical efficiency to be had from matching, other considerations argue against using a matched design. For example, much effort and expense in a case-control study is devoted to recruiting controls. Potential controls who would be suitable for an unmatched case-control study will be excluded from a matched case-control study if they fail to meet matching criteria. This reduces the sample size of a matched case-control study compared to a corresponding unmatched study, and this can inflate the variance.

On the other hand, a matched case-control study is the only feasible option when there are few cases and many confounders. Outside of this extreme situation, a blanket statement cannot be made about the choice between matched and unmatched designs. Some have argued that matching is rarely, if ever, justified in case-control studies of chronic diseases.[22] Others have recommended that matching on a limited number of strong confounders is a reasonable approach.[23]

Ratio of Controls to Cases

When designing a matched case-control study, there is little advantage to increasing the ratio of controls to cases beyond 4 : 1, which is the same as for an unmatched case-control study. See Section 12.5.

Example: Vaginal Cancer Study

Clear cell carcinoma of the vagina is a rare form of cancer that is almost unheard of in young women. During 1966–1969, eight women 15–22 years old were diagnosed with this malignancy at two Boston hospitals.[24] A case-control study was conducted in which four hospital controls were matched to each case on date of birth. The potential risk factors that were investigated in the mothers of cases included "maternal age at the birth of the child, maternal smoking (at least 10 cigarettes per day before the birth of the child), bleeding during study pregnancy, any prior pregnancy loss, maternal estrogen therapy during study pregnancy, breast feeding of infant and intrauterine X-ray exposure". Diethylstilbestrol (DES) is a synthetic form of estrogen that used to be prescribed to pregnant women for the prevention of miscarriage. Seven of the eight mothers of cases had taken DES in the first trimester of pregnancy, but none of the 32 mothers of controls had done so. Based on the findings of this and related studies, DES was withdrawn from the market for use in pregnancy. ◊

Table 13.3(a) Data for closed cohort study #7.

Exposure	Cases	Noncases	Total
$E=1$	40	60	100
$E=0$	20	80	100
Overall	60	140	200

Table 13.3(b) Data for cumulative case-control study embedded in closed cohort study #7: sampling fraction for controls 50%.

Exposure	Cases	Controls
$E=1$	40	30
$E=0$	20	40
Overall	60	70

Table 13.3(c) Misclassified data for cases in cumulative case-control study embedded in closed cohort study #7: sensitivity 80% and specificity 90%.

	$E=1$	$E=0$	Total
$E^{\star}=1$	32	2	34
$E^{\star}=0$	8	18	26
Total	40	20	60

Table 13.3(d) Misclassified data for controls in cumulative case-control study embedded in closed cohort study #7: sensitivity 80% and specificity 90%.

	$E = 1$	$E = 0$	Total
$E^{\star} = 1$	24	4	28
$E^{\star} = 0$	6	36	42
Total	30	40	70

Table 13.3(e) Misclassified data for cumulative case-control study embedded in closed cohort study #7: sensitivity 80% and specificity 90%.

Exposure	Case	Controls
$E^{\star} = 1$	34	28
$E^{\star} = 0$	26	42
Overall	60	70

Table 13.4(a) Data for case-cohort study embedded in closed cohort study #4.

Stratum	Cases		Subcohort	
	$E=1$	$E=0$	$E=1$	$E=0$
$F=1$	1,600	40	800	200
$F=0$	200	80	200	800
Overall	1,800	120	1,000	1,000

$800 = 10\% \times 8,000 \qquad 200 = 10\% \times 2,000$

$200 = 10\% \times 2,000 \qquad 800 = 10\% \times 8,000$

Table 13.4(b) Odds ratios for case-cohort study embedded in closed cohort study #4.

Crude	Stratum-specific		Stand.
	$F=1$	$F=0$	
15.0	10.0	10.0	10.0

Table 13.4(c) Classical confounding in case-cohort study embedded in closed cohort study #4.

Stratum	$\widetilde{b}_{1i}/\widetilde{b}_1$	$\widetilde{b}_{0i}/\widetilde{b}_0$	$\dfrac{\widetilde{a}_{0i}/\widetilde{b}_{0i}}{\widetilde{a}_{00}/\widetilde{b}_{00}}$
$F=1$	$0.80 = \dfrac{800}{1{,}000}$	$0.20 = \dfrac{200}{1{,}000}$	$2.0 = \dfrac{40/200}{80/800}$
$F=0$	$0.20 = \dfrac{200}{1{,}000}$	$0.80 = \dfrac{800}{1{,}000}$	$1.0 = \dfrac{80/800}{80/800}$

Table 13.5(a) Data for frequency-matched case-cohort study embedded in closed cohort study #4.

Stratum	Cases			Subcohort		
	$E=1$	$E=0$	Total	$E=1$	$E=0$	Total
$F=1$	1,600	40	1,640	1,312	328	1,640
$F=0$	200	80	280	56	224	280
Overall	1,800	120	1,920	1,368	552	1,920

$$1,312 = \frac{8,000}{8,000 + 2,000} \times 1,640 \qquad 328 = \frac{2,000}{8,000 + 2,000} \times 1,640$$

$$56 = \frac{2,000}{2,000 + 8,000} \times 280 \qquad 224 = \frac{8,000}{2,000 + 8,000} \times 280$$

Table 13.5(b) Odds ratios for frequency-matched case-cohort study embedded in closed cohort study #4.

Crude	Stratum-specific		Stand.
	$F=1$	$F=0$	
6.1	10.0	10.0	10.0

Table 13.6(a) Data and risks for closed cohort study #8.

Stratum	$E=1$			$E=0$		
	Cases	Persons	Risk[a]	Cases	Persons	Risk[a]
$F=1$	200	2,000	10.0	80	8,000	1.0
$F=0$	1,800	18,000	10.0	20	2,000	1.0
Overall	2,000	20,000	10.0	100	10,000	1.0

[a] *per 100.*

Table 13.6(b) Distribution of F at start of follow-up (percent) for closed cohort study #8.

Stratum	Exposed	Unexposed
$F=1$	10.0	80.0
$F=0$	90.0	20.0

Table 13.6(c) Measures of effect for closed cohort study #8: E is exposure and F is stratification variable.

Measure of effect	Crude	Stratum-specific		Stand.
		$F=1$	$F=0$	
Risk ratio	10.0	10.0	10.0	10.0
Odds ratio	11.0	11.0	11.0	11.0
Risk difference[a]	9.0	9.0	9.0	9.0

[a] *per 100.*

Table 13.7(a) Data for frequency-matched case-cohort study embedded in closed cohort study #8.

Stratum	Cases			Subcohort		
	$E=1$	$E=0$	Total	$E=1$	$E=0$	Total
$F=1$	200	80	280	56	224	280
$F=0$	1,800	20	1,820	1,638	182	1,820
Overall	2,000	100	2,100	1,694	406	2,100

$$56 = \frac{2,000}{2,000 + 8,000} \times 280 \qquad 224 = \frac{8,000}{2,000 + 8,000} \times 280$$

$$1,638 = \frac{18,000}{18,000 + 2,000} \times 1,820 \qquad 182 = \frac{2,000}{18,000 + 2,000} \times 1,820$$

Table 13.7(b) Odds ratios for frequency-matched case-cohort study embedded in closed cohort study #8.

Crude	Stratum-specific		Stand.
	$F=1$	$F=0$	
4.8	10.0	10.0	10.0

References

1. Berkson J. Limitations of the application of fourfold table analysis to hospital data. *Biometrics* 1946;**2**(3):47–53.
2. Boyd AV. Testing for association of diseases. *J Chron Dis* 1979;**32**(9–10):667–72.
3. Feinstein AR, Walter SD, Horwitz RI. An analysis of Berkson's bias in case-control studies. *J Chron Dis* 1986;**39**(7):495–504.
4. Flanders WD, Boyle CA, Boring JR. Bias associated with differential hospitalization rates in incident case-control studies. *J Clin Epidemiol* 1989;**42**(5):395–401.
5. Pierce N, Richiardi L. Commentary: Three worlds collide: Berkson's bias, selection bias and collider bias. *Int J Epidemiol* 2014;**43**(2):521–4.
6. Snoep JD, Moravia A, Hernández-Díaz S, et al. Commentary: A structural approach to Berkson's fallacy and a guide to a history of opinions about it. *Int J Epidemiol* 2014;**43**(2):515–21.
7. Walter SD. Berkson's bias and its control in epidemiologic studies. *J Chron Dis* 1980;**33**(11–12):721–5.
8. Hernán MA, Hernández-Díaz S, Robins JM. A structural approach to selection bias. *Epidemiology* 2004;**15**(5):615–25.
9. Birkett NJ. Effect of nondifferential misclassification on estimates of odds ratios with multiple levels of exposure. *Am J Epidemiol* 1992;**136**(3):356–62.
10. Dosemeci M, Wacholder S, Lubin JH. Does nondifferential misclassification of exposure always bias a true effect toward the null value? *Am J Epidemiol* 1990;**132**(4):746–8.
11. Knol MJ, Vandenbroucke JP, Scott P, et al. What do case-control studies estimate? Survey of methods and assumptions in published case-control research. *Am J Epidemiol* 2008;**168**(9):1073–81.
12. Miettinen OS. Components of the crude risk ratio. *Am J Epidemiol* 1972;**96**(2):168–72.
13. Mantel N, Haenszel W. Statistical aspects of the analysis of data from retrospective studies of disease. *J Natl Cancer Inst* 1959;**22**(4):719–48.
14. Lash TL, VanderWeele TJ, Haneuse S, et al., editors. *Modern epidemiology*. 4th ed. Philadelphia: Wolters Kluwer; 2021.
15. Rothman KJ. *Epidemiology: an introduction*. 2nd ed. New York: Oxford University Press; 2012.
16. Newman SC. Causal analysis of case-control data. *Epidemiol Perspect Innov* 2006;**3**:2. https://doi.org/10.1186/1742-5573-3-2 [published 27 Jan 2006].
17. Breslow NE, Day NE. *Statistical methods in cancer research. Volume 1. The analysis of case-control studies. IARC Scientific Publication number 3*. Lyon: International Agency for Research on Cancer; 1980.
18. Jewell NP. *Statistics for epidemiology*. Boca Raton: Chapman & Hall/CRC; 2004.
19. Penning de Vries BBL, Groenwold RHH. Identification of causal effects in case-control studies. *BMC Med Res Methodol* 2022;**22**:7. https://doi.org/10.1186/s12874-021-01484-7 [published 7 Jan 2022].
20. Mansournia MA, Hernán MA, Greenland S. Matched designs and causal diagrams. *Int J Epidemiol* 2013;**42**(3):860–9.
21. Thomas DC, Greenland S. The relative efficiencies of matched and independent sample designs for case-control studies. *J Chron Dis* 1983;**36**(10):685–97.
22. Howe GR, Choi BCK. Methodological issues in case-control studies: validity and power of various design/analysis strategies. *Int J Epidemiol* 1983;**12**(2):238–45.
23. Mansournia MA, Jewell NP, Greenland S. Case-control matching: effects, misconceptions, and recommendations. *Eur J Epidemiol* 2018;**33**(1):5–14.

24. Herbst AL, Ulfelder H, Poskanzer DC. Adenocarcinoma of the vagina: association of maternal stilbestrol therapy with tumor appearance in young women. *N Engl J Med* 1971;**284**(16):878–81.

CHAPTER

Ecologic Studies

14

In cohort and case-control studies, data are collected on individuals and analyzed at the individual level. By contrast, in *ecologic studies*, data are gathered on groups of individuals, usually geographically-defined populations, and analyzed at the group level.[1] Looking back at the Farr cholera study discussed in Section 1.3, we now see that it was an ecologic study of London districts and cholera mortality. Here are two more examples of ecologic studies that are relevant to studies discussed earlier.

Example: Ecologic Study of Fluoride and Caries

Two of the investigations leading up to the Newburgh-Kingston fluoride trial discussed in Section 1.10 were ecologic studies of the association between fluoride levels in community drinking water and dental health in children. Together they involved 7,257 white children 12–14 years old living in 21 communities in the United States during 1939–1940.[2,3] Dental health was measured by the same DMF scale used later in the Newburgh-Kingston fluoride trial. For each community, the fluoride concentration of the water supply was measured and the mean DMF count was determined by dental examinations. Fig. 14.1 is the corresponding scatter plot and shows a clear trend of decreasing mean DMF count with increasing fluoride concentration. ◊

Example: Ecologic Study of Birth Weight and Infant Mortality

It was pointed out in Section 6.5 as part of our discussion of the birth weight paradox that low birth weight is associated with increased infant mortality. The point prevalence of low birth weight and the infant mortality rate for each state in the United States in 2020 (excluding Vermont) were obtained from the Centers for Disease Control and Prevention/National Center for Health Statistics website.[4] Fig. 14.2 is the corresponding scatter plot and reveals a pattern of increasing infant mortality rate with increasing point prevalence of low birth weight, but the pattern is far less evident than in the preceding example. ◊

Ecologic studies are appealing because they are intuitive and often use available data sources, making them inexpensive and relatively quick to complete. There are three types of variables that appear in ecologic studies. The first type consists of summaries of measurements taken at the individual level, for example, the mean DMF count. The second type consists of measurements obtained at the group level that have a counterpart at the individual level, for example, the fluoride concentration of

Epidemiologic Methods. https://doi.org/10.1016/B978-0-44-318780-3.00020-8
Copyright © 2023 Elsevier Inc. All rights reserved.

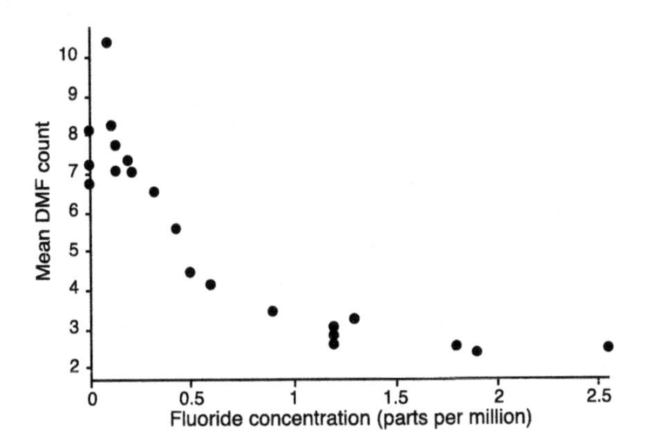

FIGURE 14.1

Mean DMF count and fluoride concentration in 21 communities of United States, 1939–1940.
Data from References 2 and 3.

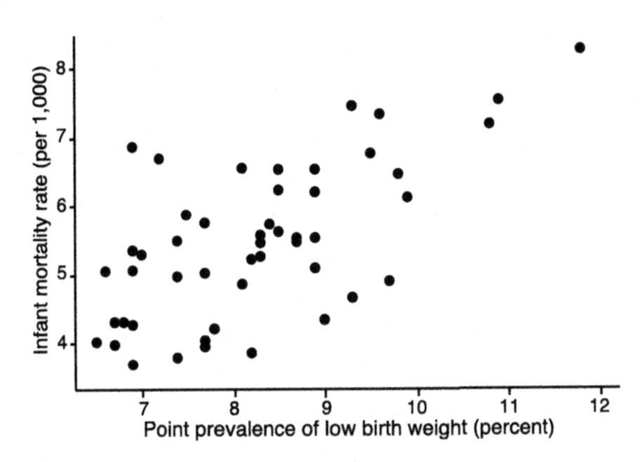

FIGURE 14.2

Infant mortality rate and point prevalence of low birth weight in states of United States (excluding Vermont), 2020.
Data from Reference 4.

the drinking water supply. The third type consists of measurements taken at the group level for which there is no counterpart at the individual level, for example, the percent of gross domestic product spent on health care.

14.1 Ecologic Bias

Although data for an ecologic study are collected at the group level, the aim of such a study in the epidemiologic setting is usually, but not always, to investigate exposure-disease associations at the individual level. This use of ecologic studies is our focus in what follows. The question then arises as to whether ecologic data are suitable for this purpose. Let us explore this fundamental issue with the help of a mathematical model.[5] It is convenient to think of the groups as finite populations and adopt the notation and terminology of Section 2.5.

Suppose there are k such populations for the ecologic study, indexed by $h = 1, \ldots, k$, with N_h individuals in population h, indexed by $i = 1, \ldots, N_h$. Let x_{hi} and y_{hi} be the exposure and disease measurements, respectively, for individual i in population h. Usually x_{hi} and y_{hi} are continuous variables, but not always. The population means of x_{hi} and y_{hi} for population h are

$$\overline{x}_h = \frac{1}{N_h} \sum_{i=1}^{N_h} x_{hi}$$
$$\overline{y}_h = \frac{1}{N_h} \sum_{i=1}^{N_h} y_{hi} \,. \qquad (14.1)$$

Table 14.1 gives the data for the ecologic study.

Table 14.1 Data for ecologic study.

Population	Exposure	Disease
1	$\overline{x}_1$	$\overline{y}_1$
$\vdots$	$\vdots$	$\vdots$
h	$\overline{x}_h$	$\overline{y}_h$
$\vdots$	$\vdots$	$\vdots$
k	$\overline{x}_k$	$\overline{y}_k$

Let us assume that within population h, but unknown to the investigators, y_{hi} is related to x_{hi} as follows:

$$y_{hi} = a_h + b x_{hi} \,, \qquad (14.2)$$

where a_h and b are unknown constants. That is, individuals in population 1 satisfy $y_{1i} = a_1 + b x_{1i}$, individuals in population 2 satisfy $y_{2i} = a_2 + b x_{2i}$, and so on. We interpret a_h as the baseline disease measurement in population h, and view b as quantifying the exposure-disease association at the individual level. For example, y_{hi} could be the risk of becoming a case for individual i in population h, and x_{hi} could be a dichotomous variable with values 0, 1 indicating exposure status. Then a_h is the risk of becoming a case in population h in the absence of exposure, and b is the risk difference for population h (and the other populations). At the individual level, b is the parameter of epidemiologic interest. Values of a_h and b could be estimated by conducting a separate study on individuals in population h, but for the ecologic study we are restricted to the population-level data in Table 14.1 and whatever analyses we can perform that might shed light on the value of b.

In addition to (14.2), let us also assume that for population h, but unknown to the investigators, a_h is related to $\overline{x}_h$ as follows:

$$a_h = a + c\overline{x}_h \,, \tag{14.3}$$

where a and c are unknown constants. We interpret a as the baseline disease measurement across all populations, and view c as quantifying the exposure-disease association at the population level.

Substituting (14.3) into (14.2) gives

$$y_{hi} = a + bx_{hi} + c\overline{x}_h \,. \tag{14.4}$$

Equation (14.4) expresses the individual-level disease measurement y_{hi} as a function of the individual-level exposure measurement x_{hi} and the population-level exposure measurement $\overline{x}_h$. This differs from what is usually encountered in epidemiology in that a group-level variable appears alongside an individual-level variable. We refer to (14.2)–(14.4) as a *multilevel model*.[5,6]

It is demonstrated in Appendix L that

$$\overline{y}_h = a + (b + c)\overline{x}_h \,. \tag{14.5}$$

This shows that according to the above model, $b + c$ is the parameter that can be estimated in an ecologic study. Although an ecologic parameter such as $b + c$ may be of interest, here our concern is the individual-level parameter b. If $c \neq 0$, that is, if there is an exposure-disease association at the population level, using $b + c$ to make inferences at the individual level results in a type of bias called *ecologic bias*, also known as the *ecologic fallacy*.[5,7,8]

As a hypothetical illustration of ecologic bias, imagine a number of countries where the probability that someone will live another year is positively correlated with individual income. Suppose that in countries with a low mean income, there is little income disparity, but as mean income increases, a disproportionate share of the additional income goes to those who already have a high income, and to such an extent that the proportion of the population with low individual income is greater in countries with high mean income than it is in countries with low mean income. In that situation, the proportion of the population of a country that will live another year will be negatively correlated with the mean income of the population.

Ecologic bias is not the only methodologic issue presented by ecologic studies. Confounding, misclassification and other sources of bias are also of concern, and can be particularly challenging in the ecologic setting.[1,8]

References

1. Morgenstern H. Ecologic studies in epidemiology: concepts, principles, and methods. *Annu Rev Public Health* 1995;**16**:61–81.
2. Dean HT, Arnold Jr FA, Elvove E. Domestic water and dental caries, V: additional studies of the relation of fluoride domestic waters to dental caries experience in 4,425 white children, aged 12 to 14 years, of 13 cities in 4 states. *Public Health Rep* 1942;**57**(32):1155–79.
3. Dean HT, Jay P, Arnold Jr FA, et al. Domestic water and dental caries, II: a study of 2,832 white children, aged 12–14 years, of 8 suburban Chicago communities, including *Lactobacillus acidophilis* studies of 1,761 children. *Public Health Rep* 1941;**56**(15):761–92.
4. Centers for Disease Control and Prevention/National Center for Health Statistics website https://www.cdc.gov/nchs. [Accessed May 2022].
5. Greenland S. A review of multilevel theory for ecologic analyses. *Stat Med* 2002;**21**(3):389–95.
6. Goldstein H. *Multilevel statistical models*. 4th ed. Chichester: Wiley; 2011.
7. Firebaugh G. A rule for inferring individual-level relationships from aggregate data. *Am Sociol Rev* 1978;**43**(4):557–72.
8. Greenland S. Ecologic versus individual-level sources of bias in ecologic estimates of contextual health effects. *Int J Epidemiol* 2001;**30**(6):1343–50.

CHAPTER

Screening in Populations 15

15.1 Natural History of Disease

Screening in the medical context can be thought of as any information-gathering activity with the aim of elucidating whether someone has a particular health condition. This happens routinely in the clinical setting when a patient presents with symptoms to a physician or other health care practitioner. Once the complaint is described, and if further investigation is warranted, the patient undergoes a *diagnostic assessment* to identify the condition causing the symptoms. In this chapter we are concerned with a radically different situation, where the goal is to determine whether someone without symptoms has an as-yet undiagnosed disease. More specifically, we are interested in population screening programs in which asymptomatic individuals are administered a *screening test* to identify those who might have an early stage of a disease. If the screening test suggests a problem, a diagnostic assessment is then performed to obtain a definitive answer. For example, in a screening program for breast cancer, the screening test might be a mammogram and the diagnostic assessment would usually include a tissue biopsy. The rationale for administering a screening test to an asymptomatic individual is that by intervening early in the natural history of the disease, the deleterious impact on health that might otherwise result will perhaps be mitigated.

The natural history of disease was discussed in Section 2.1. Fig. 15.1 is a modified version of Fig. 2.1 in which a screening test has been inserted prior to the onset of symptoms.[1–3] Note that the screening test is shown taking place during the detectable preclinical phase (DPCP). We assumed in Section 2.1 that diagnostic assessments are conducted without error and that case status at any given moment is completely determined by whether there has been a diagnostic assessment. The trigger for performing a diagnostic assessment is either a positive screening test or the onset of symptoms. We assume that in either situation the diagnostic assessment is performed promptly. By definition, a diagnostic assessment conducted on someone who is in the DPCP will identify that individual as a case. For someone in the DPCP, the time between when that person would be diagnosed following a positive screening test and when that individual would be diagnosed following the onset of symptoms is called the *lead time*. It is the length of time that the diagnosis would be advanced by screening.

Epidemiologic Methods. https://doi.org/10.1016/B978-0-44-318780-3.00021-X
Copyright © 2023 Elsevier Inc. All rights reserved.

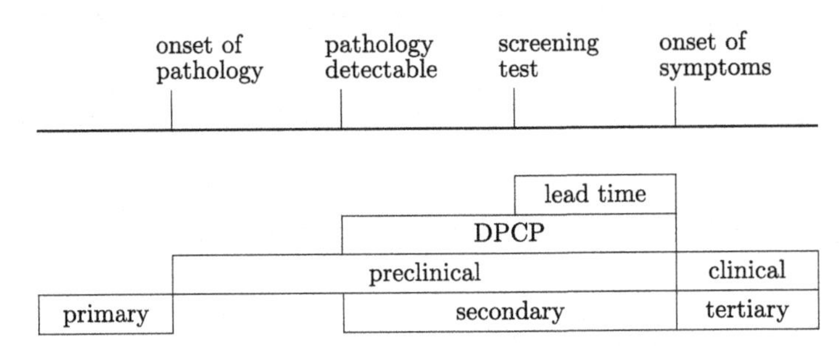

FIGURE 15.1

Natural history of disease.

15.2 Screening Programs

Screening programs can be formally organized on a community-wide basis, but usually it is left to individual health care practitioners to offer screening to their patients in accordance with accepted clinical practice. Screening guidelines are often produced by the scientific arms of specialty medical societies and by government panels such as the Canadian Task Force on Preventive Health Care (ongoing since 1976) and the U.S. Preventive Services Task Force (ongoing since 1984).

An asymptomatic individual who is diagnosed with the disease as a result of being screened is called a *screen-detected case*. Someone with symptoms who is diagnosed with the disease through the usual health care channels is said to be a *clinic-detected case*. There are two other types of cases that we need to consider—overdiagnosed cases and interval cases.

Sometimes asymptomatic disease does not evolve to the point of becoming symptomatic; that is, disease in the preclinical phase may not progress to the clinical phase. This could happen because the disease resolves during the preclinical phase, or it progresses so slowly that the individual dies of some other cause before the disease becomes symptomatic. We refer to a screen-detected case that never becomes symptomatic as an *overdiagnosed case* and say that *overdiagnosis* has occurred.[1–5] There are no advantages to being an overdiagnosed case and possibly several disadvantages. In particular, there are the unnecessary diagnostic process and the subsequent medical interventions with all their potential harms. Unfortunately, it is not possible at the present stage of medical knowledge to identify which screen-detected cases are overdiagnosed. All we can say is that in a given screening program, they could very well exist. For example, there is evidence of overdiagnosis of breast cancer by screening with mammography, and overdiagnosis of prostate cancer by screening with prostate-specific antigen.[6]

For some diseases, such as phenylketonuria in newborns, a screening program consists of administering the screening test on one occasion. For other diseases, in particular various forms of cancer, the screening program involves periodic testing at regular intervals. A participant in the latter type of screening program who is diagnosed as a case at some point in the interval between screening tests is referred to as an *interval case*.[1,2] These individuals come to medical attention in the same way, and for the same reasons, as those not participating in the screening program. The extent to which interval cases arise in a screening program depends in large part on the natural history of the disease and the length of time between screening tests.

Screening programs can be placed in the larger context of prevention in public health. There are three level of prevention, determined by their timing relative to the natural history of disease, as depicted in Fig. 15.1. *Primary prevention* is undertaken in the period prior to the onset of pathology. It has the goal of preventing the disease from occurring by removing its causes. Examples of primary prevention are seat belt legislation and anti-smoking campaigns. *Secondary prevention* takes place during the DPCP. It is aimed at detecting disease when it is asymptomatic and treatment has the potential to slow or even halt its progression. Examples of secondary prevention are screening programs for breast cancer using mammography, and for colorectal cancer using fecal occult blood testing. *Tertiary prevention* is initiated at the start of the clinical phase. It is essentially clinical care of a patient aimed at mitigating the health consequences of an illness.

Not every disease-screening test combination is suitable for a community screening program. Here are some of the criteria that the disease and screening test should satisfy.[2,3, 7(p.1433–1452)]

1. The disease must have a long enough DPCP to offer sufficient opportunity for early detection.
2. The disease must be such that early treatment is available and brings with it a potential improvement in outcome compared to later treatment.
3. The disease must be sufficiently widespread in the community and have serious enough consequences to warrant the financial and other costs associated with a screening program.
4. The screening test must cause little discomfort, produce few side-effects and be otherwise acceptable to individuals who are asymptomatic.
5. The screening test must be sufficiently valid and reliable to justify its use in a screening program.

In the next two sections we look more closely at the validity and reliability of screening tests.

15.3 Validity of Screening Tests

Prevalence studies and screening programs have certain similarities. Both are directed at a specific "population" and rely on a "test" to identify individuals who may have a "disease". A crucial difference is that a prevalence study is (usually) concerned with symptomatic disease (that is, disease in the clinical phase), whereas as a screening program aims to uncover asymptomatic disease (that is, disease in the DPCP).[1–3]

In what follows, we assume that the screening program is broadly-based enough that screened individuals are representative of the asymptomatic population. This allows us to utilize the theory developed for prevalence studies in Chapter 3. We adopt the notation established there and proceed as if the entire eligible population is involved in the screening program. Corresponding to Table 3.1, we now have Table 15.1 and the associated definitions of point prevalence, sensitivity and specificity. For present purposes, we assume that Table 15.1 represents the situation for a screening program based on a single screening episode, or the first screening episode in a periodic screening program.

Despite the similarity of the notation in Tables 3.1 and 15.1, there are conceptual differences. In Table 3.1, the focus is on case status, with D indicating true case status based on a diagnostic assessment, and $D^\star$ indicating presumed cases status based on a test. In Table 15.1, on the other hand, the focus is on DPCP status, with D indicating true DPCP status (which can only be determined by a diagnostic assessment), and $D^\star$ indicating presumed DPCP status based on a screening test. In the notation of Table 15.1, only the $A_1 + B_1$ subjects with a positive screening test undergo a diagnostic assessment as part of the screening process. Let us note that, by definition, an overdiagnosed case is counted in the true positive category, not the false positive category.

Table 15.1 Population counts for screening program.

Test	$D = 1$	$D = 0$	Total
$D^\star = 1$	A_1 (true positive)	B_1 (false positive)	$A_1 + B_1$
$D^\star = 0$	A_0 (false negative)	B_0 (true negative)	$A_0 + B_0$
Total	$A_1 + A_0$	$B_1 + B_0$	N

$$P = \frac{A_1 + A_0}{N}$$

$$S = \frac{A_1}{A_1 + A_0} \qquad (15.1)$$

$$T = \frac{B_0}{B_1 + B_0}.$$

Table 15.2 gives the sensitivities and specificities of several disease-screening test combinations currently seen in screening programs.[8–12]

In most prevalence studies, subjects are aware of their case status when they enroll, and so becoming involved in a prevalence study has few, if any, health implications. Not so for screening programs. Participants in a screening program with a true negative test benefit from the screening program because they are correctly reassured that they do not have the disease (more precisely, disease in the DPCP). Other than overdiagnosed cases, those with a true positive test benefit from the screening program if their disease is detected early and treatment at that stage improves prog-

Table 15.2 Sensitivity and specificity (percent) of selected screening programs.

Disease	Screening test	Sensitivity	Specificity
AAA[a]	ultrasound	94–100	98–100
Breast cancer	mammogram	77–95	94–97
Colorectal cancer	fecal occult blood test	62–79	87–96
Lung cancer	computed tomogram	94	73
Tuberculosis	tuberculin skin test	79	97

[a] *AAA: abdominal aortic aneurysm.*

nosis. Participants with a false positive test or a false negative test are disadvantaged by the screening program, the former because they may undergo treatment for nonexistent disease, and the latter because treatment for actual disease is delayed.

In Section 3.2 we described a straightforward method of estimating the sensitivity and specificity of a test used in prevalence studies. The corresponding problem for screening programs is more complicated because we are dealing with asymptomatic disease. It may be possible to obtain a rough estimate of the specificity from a screening program, as we now show.[1,2] As part of the screening program, the $A_1 + B_1$ individuals with a positive screening test are referred for a diagnostic assessment: A_1 of them are told they have the disease and are offered treatment; B_1 of them are informed that no disease was detected. If the point prevalence of asymptomatic disease in the community is low, then N is approximately equal to $B_1 + B_0$. It follows from (15.1) that

$$T = 1 - \frac{B_1}{B_1 + B_0} \approx 1 - \frac{B_1}{N}.$$

An estimate of the sensitivity can be obtained from a screening program that has periodic testing, but it requires an assumption that is usually difficult to defend.[2,13] As remarked above, after the first screening test, A_1 is known. Suppose the $A_0 + B_0$ individuals with a negative screening test are followed until their next screening test and that interval cases are counted. If we assume that interval cases are precisely the individuals with a false negative test, then A_0 equals the number of interval cases. Using A_1 and A_0, the sensitivity can be estimated. Alternative approaches to estimating the sensitivity based on more elaborate mathematical models are available.[13,14]

We now describe two indices that are used to quantify the diagnostic accuracy of a screening test. The *predictive value positive* of a screening test, denoted by PV_P, is the proportion of subjects with a positive test who are in the DPCP; and the *predictive value negative* of a screening test, denoted by PV_N, is the proportion of subjects with a negative test who are not in the DPCP:

$$PV_P = \frac{A_1}{A_1 + B_1}$$
$$PV_N = \frac{B_0}{A_0 + B_0}. \quad (15.2)$$

Example: Hypothetical Screening Program

Consider the hypothetical data in Table 3.2, which we now view as coming from a screening program. The point prevalence is 10%, and the sensitivity and specificity of the screening test are 80% and 90%, respectively. The predictive values are

$$PV_P = \frac{80}{170} = 47\% \quad \text{and} \quad PV_N = \frac{810}{830} = 98\%.$$

Now suppose the point prevalence is 20%, with the sensitivity and specificity unchanged. Then the predictive values are $PV_P = 67\%$ and $PV_N = 95\%$. ◊

The important lesson from the preceding example is that, unlike the sensitivity and specificity, the predictive values depend on the point prevalence. This means that the predictive values for a screening program in one population are not necessarily applicable to a screening program in another population, even when the same screening test is used. It is demonstrated in Appendix M that PV_P and PV_N can be expressed as follows:

$$PV_P = \frac{SP}{1 - T + (S + T - 1)P}$$
$$PV_N = \frac{T(1 - P)}{T - (S + T - 1)P}. \quad (15.3)$$

Fig. 15.2 shows the predictive values as a function of the point prevalence, where the sensitivity and specificity are fixed at 80% and 90%, respectively. When the point prevalence is small, which is the usual situation in a screening program, the predictive value positive is small and highly dependent on the point prevalence, whereas the predictive value negative is uniformly close to 1 and therefore of little utility in the screening setting.

Fig. 15.3 shows the predictive values as a function of the specificity, where the point prevalence is fixed at either 1% or 5%. The sensitivity is defined to be $S = 1.5 - T$, an arbitrary constraint that is meant to reflect the usual trade-off between sensitivity and specificity. When the specificity is already relatively large, increasing it further causes a substantial improvement in the predictive value positive, but almost no change in the predictive value negative.

The findings from Figs. 15.2 and 15.3 suggest that a screening program should be directed at that segment of the population most likely to have the disease; and in the trade-off between sensitivity and specificity, giving priority to specificity over sensitivity has the potential to bring a much-needed improvement in the predictive

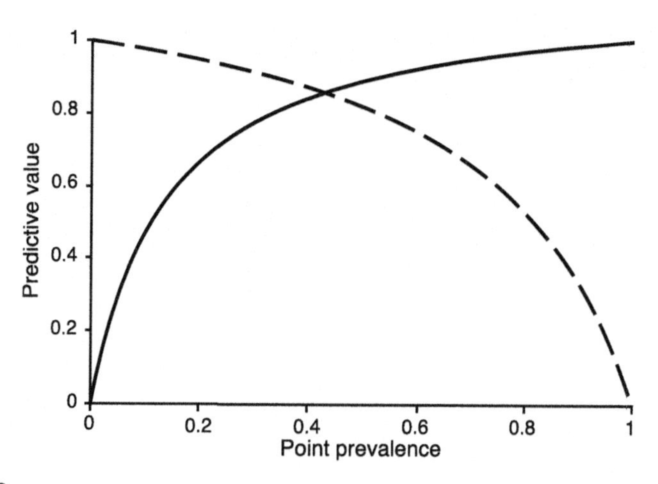

FIGURE 15.2

Positive and negative predictive value as a function of point prevalence: sensitivity 80% and specificity 90% (predictive value positive: *solid*; predictive value negative: *dash*).

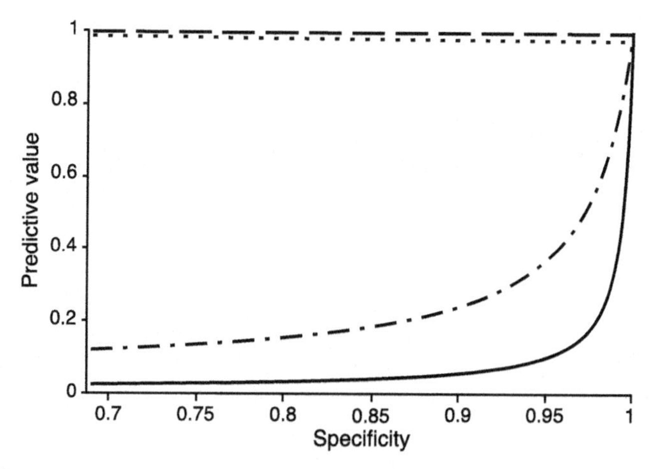

FIGURE 15.3

Positive and negative predictive value as a function of specificity, with sensitivity defined to be $S = 1.5 - T$ (predictive value negative with point prevalence of 1%: *dash*; predictive value negative with point prevalence of 5%: *dot*; predictive value positive with point prevalence of 5%: *dashdot*; predictive value positive with point prevalence of 1%: *solid*).

value positive with little impact on the predictive value negative.[1,2] A virtue of this emphasis on the predictive value positive is that it can be readily estimated from screening data because A_1 and B_1 become known.

15.4 Reliability of Screening Tests

Let us refer to the person or process responsible for interpreting a screening test as a *rater*. For example, a rater could be a trained interviewer scoring a questionnaire or a piece of laboratory equipment performing measurements on a blood sample. Reliability is concerned with the degree to which screening test results are reproducible when the screening test is administered repeatedly to the same individual, either by the same rater at different times (*intrarater reliability*) or by different raters at the same time (*interrater reliability*).

Consider a *reliability study* in which a screening test with a dichotomous outcome (positive, negative) is administered to n asymptomatic individuals by two raters, referred to as rater 1 and rater 2. For example, the subjects could be a sample of women undergoing mammography and the raters could be two radiologists. We assume that the raters make their decisions independently and that each rater has an (unknown) sensitivity and specificity relative to a diagnostic assessment (gold standard). Table 15.3 gives the rater assignments for a reliability study.

Table 15.3 Rater assignments for reliability study.

Rater 1	Rater 2		Total
	$D^\star = 1$	$D^\star = 0$	
$D^\star = 1$	a_{11}	a_{10}	$a_{11} + a_{10}$
$D^\star = 0$	a_{01}	a_{00}	$a_{01} + a_{00}$
Total	$a_{11} + a_{01}$	$a_{10} + a_{00}$	n

The raters agree that a_{11} subjects have a positive screening test and a_{00} have a negative screening test. An intuitively-appealing measure of reliability is the *proportion of agreement* (or *concordance*), defined to be the proportion of subjects on whom the raters agree:

$$p_{\text{agree}} = \frac{a_{11} + a_{00}}{n}. \tag{15.4}$$

A shortcoming of the proportion of agreement as a measure of reliability is that it does not account for agreements that are due to chance. Let us denote by p_{chance} the proportion of subjects on whom the raters agree by chance. Computing "expected" cell entries as for the Pearson chi-squared statistic, we find that

$$\begin{aligned} p_{\text{chance}} &= \left[\frac{(a_{11} + a_{01})(a_{11} + a_{10})}{n} + \frac{(a_{10} + a_{00})(a_{01} + a_{00})}{n}\right] \Big/ n \\ &= \frac{(a_{11} + a_{01})(a_{11} + a_{10}) + (a_{10} + a_{00})(a_{01} + a_{00})}{n^2}. \end{aligned} \tag{15.5}$$

Then $p_{\text{agree}} - p_{\text{chance}}$ is the proportion of individuals on whom the raters agree after correcting for chance. We could use $p_{\text{agree}} - p_{\text{chance}}$ by itself as a measure of reliability, but it is convenient to divide it by a scaling factor.

For screening tests with a dichotomous outcome, the most widely-used index of reliability in epidemiology is *kappa*, which is denoted by the Greek letter κ and defined to be[15,16]

$$\kappa = \frac{p_{\text{agree}} - p_{\text{chance}}}{1 - p_{\text{chance}}} . \tag{15.6}$$

It is easily verified that two raters agree on every subject if and only if $p_{\text{agree}} = 1$. We show in Appendix M that if both the sensitivity and specificity are 1, or if both the sensitivity and specificity are 0, then $p_{\text{agree}} = 1$. This is reasonable because in the former situation, both raters agree completely with the results of diagnostic assessments, and so agree completely with each other; while in the latter situation, both raters disagree completely with the results of diagnostic assessments, and so once again agree completely with each other.

Example: Hypothetical Reliability Study

Consider a hypothetical reliability study of 1,000 asymptomatic individuals where the point prevalence is 20%; that is, 20% of subjects would be determined to be in the DPCP if a diagnostic assessment were administered. Suppose that compared to a diagnostic assessment, rater 1 has a sensitivity of 80% and specificity of 90%, and rater 2 has a sensitivity of 60% and specificity of 70%. Tables 15.4(a) and 15.4(b) give the rater assignments for the 200 individuals in the DPCP and 800 individuals not in the DPCP, respectively. For rater 1, there are 160 subjects in the DPCP with a positive test and 720 subjects not in the DPCP with a negative test. For rater 2, there are 120 subjects in the DPCP with a positive test and 560 subjects not in the DPCP with a negative test. Summing over corresponding entries in Tables 15.4(a) and 15.4(b) gives Table 15.4(c), the rater assignments in the overall reliability study. We have from Table 15.4(c) that

$$p_{\text{agree}} = \frac{120 + 520}{1,000} = 64.0\% ,$$

$$p_{\text{chance}} = \frac{(360 \times 240) + (640 \times 760)}{(1,000)^2} = 57.3\%$$

and

$$\kappa = \frac{0.640 - 0.573}{1 - 0.573} = 15.7\% .$$

Now suppose the point prevalence is 40%, with the sensitivities and specificities unchanged. Then the kappa value is $\kappa = 21.0\%$. ◊

Table 15.4(a) Rater assignments of subjects in DPCP for hypothetical reliability study.

Rater 1	Rater 2		Total
	$D^* = 1$	$D^* = 0$	
$D^* = 1$	96 = 80% × 60% × 200	64	160 = 80% × 200
$D^* = 0$	24	16	40
Total	120 = 60% × 200	80	200 = 20% × 1,000

Table 15.4(b) Rater assignments of subjects not in DPCP for hypothetical reliability study.

Rater 1	Rater 2		Total
	$D^* = 1$	$D^* = 0$	
$D^* = 1$	24	56	80
$D^* = 0$	216	504 = 90% × 70% × 800	720 = 90% × 800
Total	240	560 = 70% × 800	800 = 80% × 1,000

Table 15.4(c) Rater assignments for hypothetical reliability study.

Rater 1	Rater 2		Total
	$D^* = 1$	$D^* = 0$	
$D^* = 1$	120	120	240
$D^* = 0$	240	520	760
Total	360	640	1,000

The preceding example shows that kappa depends not only on the sensitivities and specificities of the raters but also on the point prevalence in the population. A general formulation of the example is given in Appendix M. Under the assumption that both raters have the same sensitivity and the same specificity, it is shown in Appendix M that[16]

$$\kappa = \frac{P(1-P)(S+T-1)^2}{[1-T+(S+T-1)P][T-(S+T-1)P]}. \tag{15.7}$$

Based on (15.7), Fig. 15.4 shows kappa as a function of the point prevalence for selected sensitivities and specificities. We observe that when the point prevalence is low, which is the usual situation in community-based screening programs, the kappa value is typically no greater than 60%.

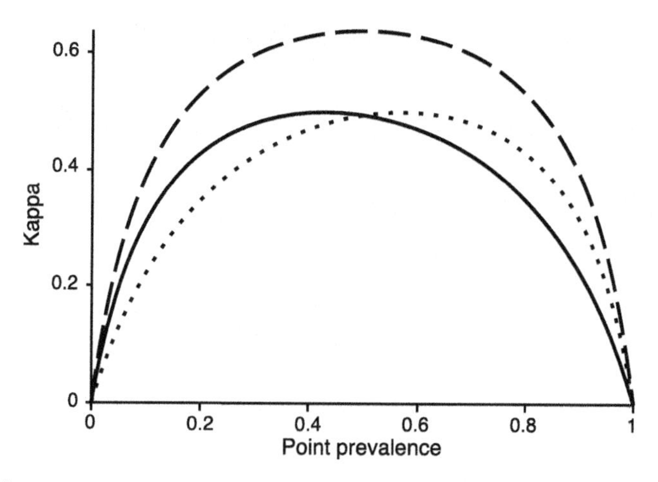

FIGURE 15.4

Kappa as a function of point prevalence: sensitivity 80% and specificity 90%: *solid*; sensitivity 90% and specificity 90%: *dash*; sensitivity 90% and specificity 80%: *dot*).

15.5 Randomized Screening Trials

Screening programs are predicated on the rationale that, all other things being equal, the earlier a diagnosis is made, the better the prognosis will be. As intuitive as this appears, before offering a screening program on a community-wide basis, there needs to be evidence that it will in fact be beneficial. A way of addressing this question is to conduct a cohort study of survival following a diagnosis of the disease in which a group of screen-detected cases is compared to a group of clinic-detected cases, with death from the disease as the study outcome. A reasonable choice of measure of effect would be the cause-specific mortality ratio (mortality rate in the screen-detected cohort divided by the mortality rate in the clinic-detected cohort). Because the disease under consideration would usually be responsible for only a small proportion of all deaths in the screen-detected and clinic-detected cohorts, screening is unlikely to have a detectable impact on the all-causes mortality ratio.

A cohort study such as the one just described is prone to four types of bias.[1–5, 7(p.1433–1452), 17,18] First, it has been observed that individuals willing to participate in health-related initiatives such as a screening program tend to have better overall health than other members of the population. When that occurs, a type of selection bias called *self-selection bias* (or *volunteer bias*) is introduced.

Second, screening programs are designed to detect disease at an earlier stage than it would be detected in the absence of screening. This results in a longer survival time as measured from the time of diagnosis until the time of death, even if the time of death is not postponed. This introduces a type of information bias called *lead time bias*.

Third, despite the appearance of Fig. 15.1, the length of the DPCP (and the length of other disease phases for that matter) varies across individuals. It is intuitively clear that the longer the DPCP, the greater the likelihood that a screening test will be performed during a part of the natural history of the disease when it can yield a positive result. It is also intuitively clear that the longer the DPCP, the slower the progression of disease. For many diseases, slower progression equates to reduced lethality. Putting these observations together, we conclude that screen-detected cases may have reduced mortality compared to clinic-detected cases, not because of any inherent benefit of the screening program, but because screening tends to identify cases with a more favorable prognosis. If so, this introduces a type of selection bias called *length bias*.

Fourth, overdiagnosed cases are not at risk of developing symptoms of the disease, let alone dying of it. Their presence in the screen-detected cohort tends to "dilute" its mortality experience, resulting in a type of selection bias called *overdiagnosis bias*.

Each of these biases lowers the cause-specific mortality rate in the cohort of screen-detected cases, making the screening program appear more beneficial than it actually is. There is an epidemiologic study design that helps to avoid these biases, namely, the randomized clinical trial, provided a measure of effect such as the cause-specific mortality ratio or cause-specific hazard ratio is used.[1–4,17,18] In the present context, we refer to such a randomized clinical trial as a *randomized screening trial*. Lead time bias is absent because all subjects begin follow-up at the time of randomization. Self-selection bias, length bias and overdiagnosis bias are potentially reduced or even absent because randomization produces balance (on average) in the screened (intervention) and nonscreened (control) groups on such factors as the intrinsic health of subjects and the length of the DPCP. Even though overdiagnosis bias is potentially eliminated in randomized screening trials, the occurrence of overdiagnosed cases in screened subjects is unaffected.

Subjects for a randomized screening trial are members of the population at risk of the disease and willing to participate in a screening program under research conditions. After signing an informed consent, subjects are randomly assigned to either the screened group or nonscreened group. At this point, there are several design options, two of which are described here. In a *continuous-screen trial*, subjects in the screened group undergo periodic screening throughout the trial, while those in the nonscreened group receive their usual medical care.[4,18] In a *stop-screen trial*, subjects in the screened group undergo periodic screening during what is called the *screening phase* of the trial, after which they receive their usual medical care during the *postscreening phase*.[4,18] Those in the nonscreened group receive their usual medical care throughout the trial. In continuous-screen and stop-screen trials, follow-up for incidence and mortality begins at the time of randomization and continues throughout the trial for both study groups. The stop-screen design is preferred when it is anticipated that a lengthy period of follow-up will be required before a reduction in mortality can be expected to emerge, which would make it expensive and logistically difficult to have periodic screening throughout the entire trial.

Imagine a stop-screen randomized screening trial in which subjects are allocated using simple randomization and where there are equal numbers of subjects in the screened and nonscreened groups. If the screening test detects the disease earlier in its natural history than it would otherwise be detected, the number of incident cases will be greater in the screened group than in the nonscreened group, at least during the screening phase and possibly for the entire trial. If the postscreening phase is at least as long as the lead time, as the trial progresses through the postscreening phase there will be more and more incident cases in the nonscreened group, gradually narrowing the gap in total incident cases between the screened and nonscreened groups. Provided the postscreening phase is long enough, a persistent difference in the number of incident cases can be attributed to overdiagnosis in the screened group.[17, 19(p.199–208)]

Example: Canadian National Breast Screening Study

The Canadian National Breast Screening Study was a randomized screening trial with a stop-screen design that investigated whether screening women with mammography decreases mortality from breast cancer.[20] The study was conducted in 15 screening centers in six Canadian provinces, with recruitment during 1980–1985. Women were eligible if they were 40–59 years old, had not had mammography in the preceding 12 months, had no history of breast cancer, and were not pregnant. Before being randomized, participants signed an informed consent. Stratified randomization was used, with stratification by study center and five-year age groups. Participants then had a physical breast examination and were taught breast self-examination.

Women 40–49 years old were randomized to receive either mammography or no mammography. Those assigned to mammography were offered another four rounds of annual mammography and physical breast examination, while those assigned to no mammography returned to their usual medical care. Women 50–59 years old were also randomized to receive either mammography or no mammography. Those assigned to mammography were offered another four rounds of annual mammography and physical breast examination, while those assigned to no mammography were offered four more rounds of annual physical breast examination without mammography. Once the screening phase was completed, women in the mammography group returned to their usual medical care. During the postscreening phase, subjects were followed for incident breast cancer and death from breast cancer using record linkage to the Canadian Cancer Registry and the Canadian National Mortality Database, respectively. For the screening and postscreening phases combined, follow-up lasted up to 25 years.

The findings for women in the 40–49 and 50–59 age groups were almost identical, and so their combined experience is presented here. There were 44,925 women in the mammography group and 44,910 in the no mammography group. The hazard ratio for death from breast cancer in the mammography group compared to the no mammography group was 0.99. There were 3,250 incident invasive breast cancers in the mammography group and 3,133 in the no mammography group. It is estimated that 106 of the women found to have breast cancers in the mammography group were overdiagnosed cases.[20] ◊

References

1. Black WC, Welch HG. Screening for disease. *Am J Roentgenol* 1997;**168**(1):3–11.
2. Cole P, Morrison AS. Basic issues in population screening for cancer. *J Natl Cancer Inst* 1980;**64**(5):1263–72.
3. Herman CR, Gill HK, Eng J, et al. Screening for preclinical disease: test and disease characteristics. *Am J Roentgenol* 2002;**179**(4):825–31.
4. Etzioni RD, Connor RJ, Prorok PC, et al. Design and analysis of cancer screening trials. *Stat Methods Med Res* 1995;**4**(1):3–17.
5. Miller AB. Conundrums in screening for cancer. *Int J Cancer* 2010;**126**(5):1039–46.
6. Welch HG, Black WC. Overdiagnosis in cancer. *J Natl Cancer Inst* 2010;**102**(9):605–13.
7. Miller AB. Fundamental issues in screening for cancer. In: Schottenfeld D, Fraumeni Jr JF, editors. *Cancer epidemiology and prevention*. 2nd ed. New York: Oxford University Press; 1996.
8. U.S. Preventive Services Task Force. Screening for abdominal aortic aneurysm: U.S. Preventive Services Task Force recommendation statement. *JAMA* 2019;**322**(22):2211–8.
9. Siu AL on behalf of the U.S. Preventive Services Task Force. Screening for breast cancer: U.S. Preventive Services Task Force recommendation statement. *Ann Intern Med* 2016;**164**(4):279–96.
10. U.S. Preventive Services Task Force. Screening for colorectal cancer: U.S. Preventive Services Task Force recommendation statement. *JAMA* 2016;**315**(23):2564–75.
11. Moyer VA on behalf of the U.S. Preventive Services Task Force. Screening for lung cancer: U.S. Preventive Services Task Force recommendation statement. *Ann Intern Med* 2014;**160**(5):330–8.
12. U.S. Preventive Services Task Force. Screening for latent tuberculosis infection in adults: U.S. Preventive Services Task Force recommendation statement. *JAMA* 2016;**316**(9):962–9.
13. Day NE. Estimating the sensitivity of a screening test. *J Epidemiol Community Health* 1985;**39**(4):364–6.
14. Prorok PC, Kramer BS, Miller AB. Study designs for determining and comparing sensitivities of disease screening tests. *J Med Screen* 2015;**22**(4):213–20.
15. Cohen J. A coefficient of agreement for nominal scales. *Educ Psychol Meas* 1960;**20**(1):37–46.
16. Thompson WD, Walter SD. A reappraisal of the kappa coefficient. *J Clin Epidemiol* 1988;**41**(10):949–58.
17. Moss S. Design issues in cancer screening trials. *Stat Methods Med Res* 2010;**19**(5):451–61.
18. Connor RJ, Prorok PC. Issues in the mortality analysis of randomized controlled trials of cancer screening. *Control Clin Trials* 1994;**15**(2):81–99.
19. Miller AB. Evidence-based cancer screening. In: Miller AB, editor. *Epidemiologic studies in cancer prevention and screening*. New York: Springer; 2013.
20. Miller AB, Wall C, Baines CJ, et al. Twenty-five year follow-up for breast cancer incidence and mortality of the Canadian National Breast Cancer Screening Study: randomised screening trial. *BMJ* 2014;**348**:g366. https://doi.org/10.1136/bmj.g366 [published 11 Feb 2014].

APPENDIX

Appendix for Chapter 2

Bracket Convention

Throughout the appendices an expression enclosed in square brackets that appears to the right of a display provides a justification for that line of the display. For example, [(2.1)] in the first line of the following display refers to expression (2.1) in Chapter 2.

A.1 Proof of Item 1 of Section 2.4

We have

$$\begin{aligned} \bar{r} &= \sum_{i=1}^{n} \left(\frac{w_i}{W} \right) r_i && [(2.1)] \\ &\leq \sum_{i=1}^{n} \left(\frac{w_i}{W} \right) r_{\max} \\ &= r_{\max} \sum_{i=1}^{n} \left(\frac{w_i}{W} \right) \\ &= r_{\max} . && [(2.2)] \end{aligned}$$

The proof of $r_{\min} \leq \bar{r}$ is similar.

APPENDIX

Appendix for Chapter 3

B.1 Proof of (3.5)

Table B.1 corresponds to Table 3.2.

Table B.1 Misclassified population counts of prevalence.

	$D=1$	$D=0$	Total
$D^{\star}=1$	SPN	$(1-T)(1-P)N$	$[SP+(1-T)(1-P)]N$
$D^{\star}=0$	$(1-S)PN$	$T(1-P)N$	$[(1-S)P+T(1-P)]N$
Total	PN	$(1-P)N$	N

Then

$$\begin{aligned} P^{\star} &= \frac{[SP+(1-T)(1-P)]N}{N} \qquad [(3.4)] \\ &= SP+(1-T)(1-P) \\ &= 1-T+(S+T-1)P\,. \end{aligned}$$

APPENDIX

Appendix for Chapter 4

C.1 Proof of (4.4)–(4.6)

The proof of (4.4) is straightforward. For the proof of (4.5), we have from (4.3) that

$$OR = RR\left(\frac{1-R_0}{1-R_1}\right). \tag{C.1}$$

Then

$$\begin{aligned}
\text{association is positive} \quad & \text{if and only if} \quad R_1 > R_0 \\
& \text{if and only if} \quad RD > 0 \\
& \text{if and only if} \quad RR > 1 \\
& \text{if and only if} \quad \frac{1-R_0}{1-R_1} > 1 \\
& \text{if and only if} \quad OR > RR, \quad [\text{(C.1)}]
\end{aligned}$$

hence the association is positive if and only if

$$OR > RR > 1 \quad \text{and} \quad RD > 0.$$

The proof of (4.6) is similar to the proof of (4.5).

C.2 Proof of (4.10)

A property of the exponential function is that if x is "small", then $\exp(x) \approx 1 + x$. By assumption, $I\Delta$ is "small", and so we have from (4.7) that $S(\Delta) \approx 1 - I\Delta$, hence

$$1 - S(\Delta) \approx I\Delta. \tag{C.2}$$

Also, since $I\Delta$ is "small", the number of incident cases in the cohort is "small", and so the number of person-years experienced by the cohort is approximately $n\Delta$. It follows that $I \approx a/(n\Delta)$, hence

$$I\Delta \approx \frac{a}{n} = R_\Delta. \tag{C.3}$$

Combining (C.2) and (C.3) gives

$$1 - S(\Delta) \approx R_\Delta.$$

APPENDIX

Appendix for Chapter 5

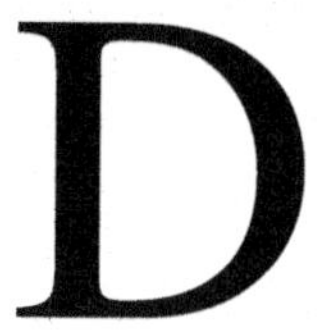

D.1 Proof of (5.8)

We have

$$\begin{aligned} I &= \frac{A}{M\Delta} && [(5.6)] \\ &= \frac{1}{M\Delta}\sum_{i=0}^{k} A_i \\ &= \frac{1}{M\Delta}\sum_{i=0}^{k} I_i M_i \Delta && [(5.7)] \\ &= \sum_{i=0}^{k}\left(\frac{M_i}{M}\right) I_i \,. \end{aligned}$$

D.2 sIR is a Weighted Average

We have

$$\begin{aligned} I_1 &= \sum_{i=0}^{k}\left(\frac{M_{1i}}{M_1}\right) I_{1i} && [(5.11)] \\ &= \frac{1}{M_1}\sum_{i=0}^{k} M_{1i} I_{0i} IR_i && [(5.9)] \end{aligned} \tag{D.1}$$

and

$$I_{0(1)} = \frac{1}{M_1}\sum_{i=0}^{k} M_{1i} I_{0i}\,, \quad [(5.16)] \tag{D.2}$$

hence

$$
\begin{aligned}
sIR &= \frac{I_1}{I_{0(1)}} && [(5.15)] \\
&= \frac{\sum_{i=0}^{k} M_{1i} I_{0i} IR_i}{\sum_{i=0}^{k} M_{1i} I_{0i}} && [(D.1), (D.2)] \\
&= \sum_{i=0}^{k} \left(\frac{M_{1i} I_{0i}}{\sum_{i=0}^{k} M_{1i} I_{0i}} \right) IR_i .
\end{aligned}
$$

D.3 Proof of (5.18)

We have

$$
I_1 M_1 \Delta = A_1 \quad [(5.6)] \tag{D.3}
$$

and

$$
\begin{aligned}
I_{0(1)} M_1 \Delta &= \sum_{i=0}^{k} I_{0i} M_{1i} \Delta && [(5.16)] \\
&= A_{0(1)} , && [(5.17)]
\end{aligned} \tag{D.4}
$$

hence

$$
\begin{aligned}
sIR &= \frac{I_1}{I_{0(1)}} && [(5.15)] \\
&= \frac{I_1 M_1 \Delta}{I_{0(1)} M_1 \Delta} \\
&= \frac{A_1}{A_{0(1)}} . && [(D.3), (D.4)]
\end{aligned}
$$

Appendix for Chapter 6

E.1 Proof of (6.6)

We have

$$\frac{RR_{E|F=1}}{RR_{E|F=0}} = \frac{R_{E=1,F=1}/R_{E=0,F=1}}{R_{E=1,F=0}/R_{E=0,F=0}} \quad [(6.1)]$$

$$= \frac{R_{E=1,F=1}/R_{E=1,F=0}}{R_{E=0,F=1}/R_{E=0,F=0}}$$

$$= \frac{RR_{F|E=1}}{RR_{F|E=0}}, \quad [(6.1)]$$

hence

$$RR_{E|F=1} = RR_{E|F=0} \quad \text{if and only if} \quad RR_{F|E=1} = RR_{F|E=0}.$$

That is, "the risk ratio for the E-D association is homogeneous over F if and only if the risk ratio for the F-D association is homogeneous over E", which is equivalent to "F is an effect modifier of the risk ratio if and only if E is an effect modifier of the risk ratio".

We also have

$$RD_{E|F=1} - RD_{E|F=0}$$

$$= (R_{E=1,F=1} - R_{E=0,F=1}) - (R_{E=1,F=0} - R_{E=0,F=0}) \quad [(6.1)]$$

$$= (R_{E=1,F=1} - R_{E=1,F=0}) - (R_{E=0,F=1} - R_{E=0,F=0})$$

$$= RD_{F|E=1} - RD_{F|E=0}, \quad [(6.1)]$$

hence

$$RD_{E|F=1} = RD_{E|F=0} \quad \text{if and only if} \quad RD_{F|E=1} = RD_{F|E=0}.$$

That is, "the risk difference for the E-D association is homogeneous over F if and only if the risk difference for the F-D association is homogeneous over E", which is equivalent to "F is an effect modifier of the risk difference if and only if E is an effect modifier of the risk difference".

E.2 Proof of (6.7)

The risk ratio for the E-D association is homogeneous over F if and only if

$$RR_{E|F=0} = RR_{E|F=1} \qquad [(6.4)]$$
$$= \frac{R_{E=1,F=1}}{R_{E=0,F=1}}, \qquad [(6.1)]$$

which is equivalent to

$$R_{E=1,F=1} = RR_{E|F=0} \times R_{E=0,F=1}$$
$$= R_{E=0,F=0} \times RR_{E|F=0} \times \frac{R_{E=0,F=1}}{R_{E=0,F=0}}$$
$$= R_{E=0,F=0} \times RR_{E|F=0} \times RR_{F|E=0}. \qquad [(6.1)]$$

E.3 Proof of (6.8)

The risk difference for the E-D association is homogeneous over F if and only if

$$RD_{E|F=0} = RD_{E|F=1} \qquad [(6.5)]$$
$$= R_{E=1,F=1} - R_{E=0,F=1}, \qquad [(6.1)]$$

which is equivalent to

$$R_{E=1,F=1} = RD_{E|F=0} + R_{E=0,F=1}$$
$$= R_{E=0,F=0} + RD_{E|F=0} + (R_{E=0,F=1} - R_{E=0,F=0})$$
$$= R_{E=0,F=0} + RD_{E|F=0} + RD_{F|E=0}. \qquad [(6.1)]$$

E.4 Proof of (6.11)

Suppose both the risk ratio and risk difference for the E-D association are homogeneous over F. By definition, this is equivalent to (6.4) and (6.5) being true. We have from (6.7) that

$$\frac{R_{E=1,F=1}}{R_{E=0,F=0}} = RR_{E|F=0} \times RR_{F|E=0}. \qquad \text{(E.1)}$$

We also have

$$
\begin{aligned}
&(R_{E=1,F=1} - R_{E=0,F=1}) - (R_{E=1,F=0} - R_{E=0,F=0}) \\
&= RD_{E|F=1} - RD_{E|F=0} \qquad [(6.1)] \\
&= 0, \qquad [(6.5)]
\end{aligned}
$$

hence

$$R_{E=1,F=1} = R_{E=0,F=1} + R_{E=1,F=0} - R_{E=0,F=0},$$

and so

$$
\begin{aligned}
\frac{R_{E=1,F=1}}{R_{E=0,F=0}} &= \frac{R_{E=0,F=1}}{R_{E=0,F=0}} + \frac{R_{E=1,F=0}}{R_{E=0,F=0}} - 1 \\
&= RR_{F|E=0} + RR_{E|F=0} - 1. \qquad [(6.1)]
\end{aligned}
\tag{E.2}
$$

Combining (E.1) and (E.2) gives

$$RR_{E|F=0} \times RR_{F|E=0} = RR_{F|E=0} + RR_{E|F=0} - 1,$$

hence

$$
\begin{aligned}
0 &= 1 - RR_{E|F=0} - RR_{F|E=0} + (RR_{E|F=0} \times RR_{F|E=0}) \\
&= (1 - RR_{E|F=0})(1 - RR_{F|E=0}).
\end{aligned}
$$

It follows that either $RR_{E|F=0} = 1$ or $RR_{F|E=0} = 1$, or both; that is, (6.9) is not true. This shows that "if both the risk ratio and risk difference for the E-D association are homogeneous over F, then (6.9) is not true", which is logically equivalent to "if (6.9) is true, then the risk ratio for the E-D association is not homogeneous over F, or the risk difference for the E-D association is not homogeneous over F, or both", which is equivalent to "if (6.9) is true, then F is an effect modifier of the risk ratio or the risk difference, or both".

To prove the "likewise for E" part of (6.11), we simply interchange the roles of E and F in the preceding argument.

E.5 Proof of (6.15)

We have

$$
\begin{aligned}
RR_{E|F=1} < 1 \quad &\text{if and only if} \quad \frac{R_{E=1,F=1}}{R_{E=0,F=1}} < 1 \qquad [(6.1)] \\
&\text{if and only if} \quad R_{E=1,F=1} < R_{E=0,F=1} \\
&\text{if and only if} \quad (R_{E=1,F=1} - R_{E=0,F=1}) < 0 \\
&\text{if and only if} \quad RD_{E|F=1} < 0, \qquad [(6.1)]
\end{aligned}
$$

and likewise,

$$RR_{E|F=0} > 1 \quad \text{if and only if} \quad RD_{E|F=0} > 0.$$

It follows that

$$RR_{E|F=1} < 1 < RR_{E|F=0} \quad \text{if and only if} \quad RD_{E|F=1} < 0 < RR_{E|F=0}.$$

Similarly,

$$RR_{E|F=1} > 1 > RR_{E|F=0} \quad \text{if and only if} \quad RD_{E|F=1} > 0 > RD_{E|F=0}.$$

The above argument also shows that when F is a qualitative effect modifier of the risk ratio or risk difference, the stratum-specific values of the risk ratio and risk difference exhibit the same trend over strata of F.

APPENDIX

Appendix for Chapter 7

F.1 sRR, sRD, RR_{MH}, OR_{MH}, RD_{MH} are Weighted Averages

We have from (7.4) that

$$
\begin{aligned}
sRR &= \frac{R_1}{R_{0(1)}} \\
&= \sum_{i=0}^{k}\left(\frac{n_{1i}}{n_1}\right)R_{1i} \Big/ \sum_{i=0}^{k}\left(\frac{n_{1i}}{n_1}\right)R_{0i} \qquad [(7.2),\ (7.3)] \\
&= \sum_{i=0}^{k}\left(\frac{n_{1i}R_{0i}}{n_1}\right)RR_i \Big/ \sum_{i=0}^{k}\frac{n_{1i}R_{0i}}{n_1} \qquad [(7.1)] \\
&= \sum_{i=0}^{k}\left(\frac{w_i}{\sum_{i=0}^{k} w_i}\right)RR_i \qquad \left[w_i = \frac{n_{1i}R_{0i}}{n_1}\right]
\end{aligned}
\tag{F.1}
$$

and

$$
\begin{aligned}
sRD &= R_1 - R_{0(1)} \\
&= \sum_{i=0}^{k}\left(\frac{n_{1i}}{n_1}\right)R_{1i} - \sum_{i=0}^{k}\left(\frac{n_{1i}}{n_1}\right)R_{0i} \qquad [(7.2),\ (7.3)] \\
&= \sum_{i=0}^{k}\left(\frac{n_{1i}}{\sum_{i=0}^{k} n_{1i}}\right)RD_i \qquad [(7.1),\ \text{Table 7.3}] \\
&= \sum_{i=0}^{k}\left(\frac{w_i}{\sum_{i=0}^{k} w_i}\right)RD_i\,, \qquad [w_i = n_{1i}]
\end{aligned}
$$

and from (7.7) that

$$RR_{\text{MH}} = \sum_{i=0}^{k} \frac{a_{1i}n_{0i}}{n_{0i}+n_{1i}} \Bigg/ \sum_{i=0}^{k} \frac{a_{0i}n_{1i}}{n_{0i}+n_{1i}}$$

$$= \sum_{i=0}^{k} \left(\frac{a_{0i}n_{1i}}{n_{0i}+n_{1i}}\right) RR_i \Bigg/ \sum_{i=0}^{k} \frac{a_{0i}n_{1i}}{n_{0i}+n_{1i}} \quad [(4.3), \text{Table 7.3}]$$

$$= \sum_{i=0}^{k} \left(\frac{w_i}{\sum_{i=0}^{k} w_i}\right) RR_i\,, \quad \left[w_i = \frac{a_{0i}n_{1i}}{n_{0i}+n_{1i}}\right]$$

$$OR_{\text{MH}} = \sum_{i=0}^{k} \frac{a_{1i}b_{0i}}{n_{0i}+n_{1i}} \Bigg/ \sum_{i=0}^{k} \frac{a_{0i}b_{1i}}{n_{0i}+n_{1i}}$$

$$= \sum_{i=0}^{k} \left(\frac{a_{0i}b_{1i}}{n_{0i}+n_{1i}}\right) OR_i \Bigg/ \sum_{i=0}^{k} \frac{a_{0i}b_{1i}}{n_{0i}+n_{1i}} \quad [(4.3), \text{Table 7.3}]$$

$$= \sum_{i=0}^{k} \left(\frac{w_i}{\sum_{i=0}^{k} w_i}\right) OR_i \quad \left[w_i = \frac{a_{0i}b_{1i}}{n_{0i}+n_{1i}}\right]$$

and

$$RD_{\text{MH}} = \sum_{i=0}^{k} \frac{a_{1i}n_{0i} - a_{0i}n_{1i}}{n_{0i}+n_{1i}} \Bigg/ \sum_{i=0}^{k} \frac{n_{0i}n_{1i}}{n_{0i}+n_{1i}}$$

$$= \sum_{i=0}^{k} \left(\frac{n_{0i}n_{1i}}{n_{0i}+n_{1i}}\right) RD_i \Bigg/ \sum_{i=0}^{k} \frac{n_{0i}n_{1i}}{n_{0i}+n_{1i}} \quad [(4.3), \text{Table 7.3}]$$

$$= \sum_{i=0}^{k} \left(\frac{w_i}{\sum_{i=0}^{k} w_i}\right) RD_i\,. \quad \left[w_i = \frac{n_{0i}n_{1i}}{n_{0i}+n_{1i}}\right]$$

F.2 sIR, sID, IR_{MH}, ID_{MH} are Weighted Averages

Computing as in Section F.1, we have from (7.12) that

$$sIR = \sum_{i=0}^{k} \left(\frac{w_i}{\sum_{i=0}^{k} w_i}\right) IR_i \quad \left[w_i = \frac{y_{1i}I_{0i}}{y_1}\right]$$

$$sID = \sum_{i=0}^{k} \left(\frac{w_i}{\sum_{i=0}^{k} w_i} \right) ID_i \quad [w_i = y_{1i}]$$

and from (7.13) that

$$IR_{\text{MH}} = \sum_{i=0}^{k} \left(\frac{w_i}{\sum_{i=0}^{k} w_i} \right) IR_i \quad \left[w_i = \frac{a_{0i} y_{1i}}{y_{0i} + y_{1i}} \right]$$

$$ID_{\text{MH}} = \sum_{i=0}^{k} \left(\frac{w_i}{\sum_{i=0}^{k} w_i} \right) ID_i \quad \left[w_i = \frac{y_{0i} y_{1i}}{y_{0i} + y_{1i}} \right].$$

F.3 Proof of (7.11)

By assumption, the risk of becoming a case is "small". Since R_{0i} and R_{1i} are "small", it follows from (7.1) that

$$OR_i \approx \frac{R_{1i}}{R_{0i}} = RR_i\,. \tag{F.2}$$

Since R_{0i} is "small", it follows from (7.3) that $R_{0(1)}$ is "small". Since R_1 and $R_{0(1)}$ are "small", it follows from (7.4) that

$$sOR \approx \frac{R_1}{R_{0(1)}} = sRR\,. \tag{F.3}$$

By assumption, the odds ratio is homogeneous, that is,

$$OR_i = hOR\,. \tag{F.4}$$

Combining (F.2) and (F.4) yields

$$RR_i \approx hOR\,. \tag{F.5}$$

It follows from item 3 of Section 2.4, the expression for sRR in (F.1), and (F.5) that

$$sRR \approx hOR\,. \tag{F.6}$$

Then (F.3) and (F.6) give

$$sOR \approx hOR\,.$$

APPENDIX

Appendix for Chapter 8

G.1 Proof of (8.2)

Consistent with (7.2), the crude risk in the exposed cohort in the absence of exposure is

$$R_1^\bullet = \sum_{i=0}^{k} \left(\frac{n_{1i}^\bullet}{n_1^\bullet} \right) R_{1i}^\bullet , \tag{G.1}$$

where $n_1^\bullet$, $n_{1i}^\bullet$ and $R_{1i}^\bullet$ are the counterfactual counterparts of n_1, n_{1i} and R_{1i}, respectively. It is clear that the exposed cohort and the exposed cohort in the absence of exposure have the same number of subjects at the start of follow-up:

$$n_1^\bullet = n_1 . \tag{G.2}$$

(C5) implies that the exposed cohort and the exposed cohort in the absence of exposure have the same distribution of F at the start of follow-up. It follows from this observation and (G.2) that the exposed cohort and the exposed cohort in the absence of exposure have same the number of subjects in each stratum of F at the start of follow-up:

$$n_{1i}^\bullet = n_{1i} . \tag{G.3}$$

Consistent with (C4), we observe that (C6) can be expressed as

$$R_{1i}^\bullet = R_{0i} . \tag{G.4}$$

Substituting (G.2)–(G.4) into (G.1) yields

$$R_1^\bullet = \sum_{i=0}^{k} \left(\frac{n_{1i}}{n_1} \right) R_{0i} = R_{0(1)} ,$$

where the second equality comes from (7.3).

G.2 Proof of (8.3)

We have

$$
\begin{aligned}
a_1^\bullet &= R_1^\bullet n_1^\bullet & [(4.1)] \\
&= R_{0(1)} n_1 & [(8.2), (G.2)] \\
&= a_{0(1)} . & [(7.5)]
\end{aligned}
$$

G.3 Proof of Item 1 of Section 8.4

We have from (7.2) and (7.3) that $R_0 = R_{0(1)}$ is equivalent to

$$\sum_{i=0}^{k} \left(\frac{n_{0i}}{n_0}\right) R_{0i} = \sum_{i=0}^{k} \left(\frac{n_{1i}}{n_1}\right) R_{0i} . \tag{G.5}$$

It follows that if $n_{1i}/n_1 = n_{0i}/n_0$ for all values of i, or if $R_{00}, \ldots, R_{0i}, \ldots, R_{0k}$ are all equal, then $R_0 = R_{0(1)}$. That is, "if (C13) or (C14) is not satisfied, then (C7) is not satisfied", which is logically equivalent to "if (C7) is satisfied, then (C13) and (C14) are satisfied". This shows that if F is a counterfactual confounder, then it is a classical confounder.

G.4 Proof of Item 2 of Section 8.4

If F is dichotomous, then $k = 1$ and (G.5) becomes

$$\left(\frac{n_{00}}{n_0}\right) R_{00} + \left(\frac{n_{01}}{n_0}\right) R_{01} = \left(\frac{n_{10}}{n_1}\right) R_{00} + \left(\frac{n_{11}}{n_1}\right) R_{01} .$$

We have from Table 7.3 that $n_{00} + n_{01} = n_0$ and $n_{10} + n_{11} = n_1$, and so $R_0 = R_{0(1)}$ is equivalent to

$$\left(\frac{n_{00}}{n_0}\right) R_{00} + \left(1 - \frac{n_{00}}{n_0}\right) R_{01} = \left(\frac{n_{10}}{n_1}\right) R_{00} + \left(1 - \frac{n_{10}}{n_1}\right) R_{01} ,$$

which is equivalent to

$$\left[\left(\frac{n_{00}}{n_0}\right) - \left(\frac{n_{10}}{n_1}\right)\right] (R_{00} - R_{01}) = 0 .$$

It follows that "$R_0 = R_{0(1)}$ if and only if $n_{10}/n_1 = n_{00}/n_0$ or $R_{00} = R_{01}$", which is logically equivalent to "$R_0 \neq R_{0(1)}$ if and only if $n_{10}/n_1 \neq n_{00}/n_0$ and $R_{00} \neq R_{01}$". That is, (C7) is satisfied if and only if (C13) and (C14) are satisfied. This shows that when F is dichotomous, it is a counterfactual confounder if and only if it is a classical confounder.

G.5 Proof of Item 1 of Section 8.5

Without loss of generality, we assume for simplicity of notation that the weights $w_0 \ldots, w_i, \ldots, w_k$ satisfy $\sum_{i=0}^{k} w_i = 1$, and that the stratum-specific measures of effect $M_0 \ldots, M_i, \ldots, M_k$ satisfy $M_{\min} = M_0$ and $M_{\max} = M_k$.

Suppose M is collapsible over F. By definition, $M = \sum_{i=0}^{k} w_i M_i$, for some weights $w_0, \ldots, w_i, \ldots, w_k$. It follows from item 1 of Section 2.4 that $M_0 \leq M \leq M_k$.

Conversely, suppose $M_0 \leq M \leq M_k$. There are three cases to consider. (1) If $M = M_0$, let $w_0 = 1$ and $w_i = 0$ for $i > 0$. (2) If $M = M_k$, let $w_k = 1$ and $w_i = 0$ for $i < k$. (3) If $M_0 < M < M_k$, let $w_0 = (M_k - M)/(M_k - M_0)$, $w_k = 1 - w_0$, and $w_i = 0$ for $0 < i < k$. In each case, $M = \sum_{i=0}^{k} w_i M_i$.

G.6 Proof of Item 2 of Section 8.5

We have from (F.1) that sRR is collapsible over F. Then

F is not a counterfactual confounder		
if and only if	$R_0 = R_{0(1)}$	[(C7)]
if and only if	$RR = sRR$	[(C8)]
implies	RR is collapsible over F.	[sRR collapsible over F]

A similar argument applies to RD.

APPENDIX

Appendix for Chapter 9

H.1 No Counterfactual Confounding Implies $a_{0(1)} = R_0 n_1$

If there is no counterfactual confounding, then

$$R_0 n_1 = R_{0(1)} n_1 \quad [(C7)]$$
$$= a_{0(1)}. \quad [(7.5)]$$

H.2 Proof of (9.7)

We have

$$sAF_E = \frac{a_1 - a_{0(1)}}{a_1} \quad [(9.1)]$$
$$= \frac{a_1/a_{0(1)} - 1}{a_1/a_{0(1)}}$$
$$= \frac{sRR - 1}{sRR}. \quad [(7.6)]$$

H.3 Proof of (9.5)

If there is no counterfactual confounding, we have from (C8) that $sRR = RR$, and then (9.7) becomes

$$AF_E = \frac{RR - 1}{RR}.$$

H.4 Proof of First Equality in (9.8)

We have

$$sAF_P = \frac{a_1 - a_{0(1)}}{a} \quad [(9.2)]$$

$$= \frac{a_1}{a} \times \frac{a_1 - a_{0(1)}}{a_1}$$

$$= f \times sAF_E. \qquad [(9.1), (9.4)]$$

H.5 Proof of Second Equality in (9.8)

We have

$$sAF_P = \frac{a_1 - a_{0(1)}}{a} \qquad [(9.2)]$$

$$= \frac{\sum_{i=0}^{k} a_{1i} - \sum_{i=0}^{k} R_{0i} n_{1i}}{a} \qquad \text{[Table 7.3, (7.5)]}$$

$$= \sum_{i=0}^{k} \frac{a_{1i} - R_{0i} n_{1i}}{a}$$

$$= \sum_{i=0}^{k} \left(\frac{a_{1i} + a_{0i}}{a} \times \frac{a_{1i} - R_{0i} n_{1i}}{a_{1i} + a_{0i}} \right)$$

$$= \sum_{i=0}^{k} h_i \times AF_{Pi}. \qquad [(9.3), (9.4)]$$

H.6 Proof of First Equality in (9.6)

Since there is no counterfactual confounding, we have

$$AF_P = \frac{a_1 - R_0 n_1}{a} \qquad [(9.3)]$$

$$= \frac{a_1}{a} \times \frac{a_1 - R_0 n_1}{a_1}$$

$$= f \times AF_E. \qquad [(9.3), (9.4)]$$

H.7 Proof of Second Equality in (9.6)

We have

$$\begin{aligned} a_1 - R_0 n_1 &= R_1 n_1 - R_0 n_1 \quad [(4.2)] \\ &= (R_1 - R_0) n_1 \end{aligned} \qquad \text{(H.1)}$$

and

$$\begin{aligned} a &= a_1 + a_0 && \text{[Table 4.1]} \\ &= R_1 n_1 + R_0 n_0 && \text{[(4.2)]} \\ &= R_1 n_1 + R_0 (n - n_1) && \text{[Table 4.1]} \\ &= (R_1 - R_0) n_1 + R_0 n\,, \end{aligned} \tag{H.2}$$

hence

$$\begin{aligned} AF_P &= \frac{a_1 - R_0 n_1}{a} && \text{[(9.3)]} \\ &= \frac{n_1(R_1 - R_0)}{n_1(R_1 - R_0) + nR_0} && \text{[(H.1), (H.2)]} \\ &= \frac{g(R_1 - R_0)}{g(R_1 - R_0) + R_0} && \text{[(9.4)]} \\ &= \frac{g(RR - 1)}{g(RR - 1) + 1}\,. && \text{[(4.3)]} \end{aligned}$$

H.8 Proof of (9.9)

In the notation of (4.2), (6.1) and Tables 7.1 and 7.3, we have

$$\begin{aligned} R_{E=1} &= \frac{a_1}{n_1} \\ &= \frac{a_{11} + a_{10}}{n_1} \\ &= \left(\frac{n_{11}}{n_1}\right)\left(\frac{a_{11}}{n_{11}}\right) + \left(\frac{n_{10}}{n_1}\right)\left(\frac{a_{10}}{n_{10}}\right) \\ &= p_{F=1|E=1} R_{E=1,F=1} + (1 - p_{F=1|E=1}) R_{E=1,F=0} \\ &= R_{E=1,F=0}\left[p_{F=1|E=1}\left(\frac{R_{E=1,F=1}}{R_{E=1,F=0}}\right) + 1 - p_{F=1|E=1}\right] \\ &= R_{E=1,F=0}\left[p_{F=1|E=1}(RR_{F|E=1} - 1) + 1\right], \end{aligned}$$

and similarly,

$$R_{E=0} = R_{E=0,F=0}\left[p_{F=1|E=0}(RR_{F|E=0} - 1) + 1\right],$$

hence

$$\begin{aligned} cRR_E &= \frac{R_{E=1}}{R_{E=0}} \\ &= \frac{R_{E=1,F=0}\,[p_{F=1|E=1}(RR_{F|E=1} - 1) + 1]}{R_{E=0,F=0}\,[p_{F=1|E=0}(RR_{F|E=0} - 1) + 1]} \\ &= RR_{E|F=0}\left[\frac{p_{F=1|E=1}(RR_{F|E=1} - 1) + 1}{p_{F=1|E=0}(RR_{F|E=0} - 1) + 1}\right]. \end{aligned} \tag{H.3}$$

It follows from the assumptions set out in Section 9.2 that

$$RR_{E|F=0} = hRR_E \tag{H.4}$$

and

$$RR_{F|E=1} = RR_{F|E=0} = hRR_F\,. \tag{H.5}$$

Substituting (H.4) and (H.5) into (H.3) and rearranging terms gives (9.9).

APPENDIX

Appendix for Chapter 10

I

I.1 Proof of (10.1)

Tables I.1(a)–I.1(d) correspond to Tables 10.1(a)–10.1(d), respectively.

Table I.1(a) Data for closed cohort study.

Exposure	$D=1$	$D=0$	Persons
$E=1$	R_1n_1	$(1-R_1)n_1$	n_1
$E=0$	R_0n_0	$(1-R_0)n_0$	n_0

Table I.1(b) Misclassified data for exposed subjects in closed cohort study.

	$D=1$	$D=0$	Total
$D^\star=1$	SR_1n_1	$(1-T)(1-R_1)n_1$	$[1-T+(S+T-1)R_1]n_1$
$D^\star=0$	$(1-S)R_1n_1$	$T(1-R_1)n_1$	$[T-(S+T-1)R_1]n_1$
Total	R_1n_1	$(1-R_1)n_1$	n_1

Table I.1(c) Misclassified data for unexposed subjects in closed cohort study.

	$D=1$	$D=0$	Total
$D^\star=1$	SR_0n_0	$(1-T)(1-R_0)n_0$	$[1-T+(S+T-1)R_0]n_0$
$D^\star=0$	$(1-S)R_0n_0$	$T(1-R_0)n_0$	$[T-(S+T-1)R_0]n_0$
Total	R_0n_0	$(1-R_0)n_0$	n_0

Table I.1(d) Misclassified data for closed cohort study.

Exposure	$D^\star = 1$	$D^\star = 0$	Persons
$E = 1$	$a_1^\star$	$b_1^\star$	n_1
$E = 0$	$a_0^\star$	$b_0^\star$	n_0

$$a_1^\star = [1 - T + (S + T - 1)R_1]n_1$$

$$a_0^\star = [1 - T + (S + T - 1)R_0]n_0$$

$$b_1^\star = [T - (S + T - 1)R_1]n_1$$

$$b_0^\star = [T - (S + T - 1)R_0]n_0 .$$

It follows from Table I.1(d) that

$$\begin{aligned} RR^\star &= \frac{a_1^\star/n_1}{a_0^\star/n_0} \\ &= \frac{1 - T + (S + T - 1)R_1}{1 - T + (S + T - 1)R_0} \\ OR^\star &= \frac{a_1^\star b_0^\star}{a_0^\star b_1^\star} \\ &= \frac{[1 - T + (S + T - 1)R_1][T - (S + T - 1)R_0]}{[1 - T + (S + T - 1)R_0][T - (S + T - 1)R_1]} \\ RD^\star &= \frac{a_1^\star}{n_1} - \frac{a_0^\star}{n_0} \\ &= [1 - T + (S + T - 1)R_1] - [1 - T + (S + T - 1)R_0] \\ &= (S + T - 1)RD . \end{aligned} \tag{I.1}$$

I.2 Proof of (10.2)

Substituting $R_1 = RR \times R_0$ into the identity for $RR^\star$ in (I.1) gives

$$RR^\star = \frac{1 - T + RR(S + T - 1)R_0}{1 - T + (S + T - 1)R_0} . \tag{I.2}$$

By assumption, $R_0 > 0$. Also by assumption, $0 < T < 1$ and $S + T > 1$, and so $1 - T > 0$ and $S + T - 1 > 0$. If $RR > 1$, then

$$1 - T + RR(S + T - 1)R_0 > 1 - T + (S + T - 1)R_0 \tag{I.3}$$

and

$$1 - T + RR(S + T - 1)R_0 < RR[1 - T + (S + T - 1)R_0]. \tag{I.4}$$

It follows from (I.2) and (I.3) that $RR^\star > 1$, and from (I.2) and (I.4) that $RR^\star < RR$. Thus, $RR > RR^\star > 1$. Similarly, if $RR < 1$, then $RR < RR^\star < 1$.

I.3 Proof of (10.3)

Expanding the identity for $OR^\star$ in (I.1) and carrying out a lengthy but straightforward computation gives

$$OR^\star = \frac{OR + U + V}{1 + (U \times OR) + V}, \tag{I.5}$$

where

$$\begin{aligned} U &= \frac{(1-S)(1-T)}{ST} \\ V &= \frac{(1-S)R_1}{T(1-R_1)} + \frac{(1-T)(1-R_0)}{SR_0}. \end{aligned} \tag{I.6}$$

Suppose $OR > 1$. By assumption, $S > 0$ and $T > 0$. Then

$$\begin{aligned} OR^\star > 1 \quad &\text{if and only if} \quad OR + U + V > 1 + (U \times OR) + V && [\text{(I.5)}] \\ &\text{if and only if} \quad (1-U)OR > 1 - U \\ &\text{if and only if} \quad U < 1 && [OR > 1] \\ &\text{if and only if} \quad \frac{(1-S)(1-T)}{ST} < 1 && [\text{(I.6)}] \\ &\text{if and only if} \quad S + T > 1. \end{aligned}$$

By assumption, $S + T > 1$, and so $OR^\star > 1$.

By assumption, $0 < S < 1$ and $0 < T < 1$, and so $1 - S > 0$ and $1 - T > 0$, hence $U > 0$. Also by assumption, $0 < R_0 < 1$ and $0 < R_1 < 1$, and so $1 - R_0 > 0$ and $1 - R_1 > 0$, hence $V > 0$. Then

$$\begin{aligned} OR^\star &< \frac{OR + U + V}{1 + U + V} && [\text{(I.5)}, OR > 1] \\ &< \frac{OR + OR(U + V)}{1 + U + V} && [OR > 1] \\ &= OR. \end{aligned}$$

Combining the inequalities $OR > OR^\star$ and $OR^\star > 1$ yields $OR > OR^\star > 1$. Similarly, if $OR < 1$, then $OR < OR^\star < 1$.

I.4 Proof of (10.4)

By assumption, $0 < S < 1$ and $0 < T < 1$ and $S + T > 1$, hence $0 < S + T - 1 < 1$. It follows from the identity for $RD^\star$ in (I.1) that if $RD > 0$, then $RD^\star > 0$ and $RD > RD^\star$, hence $RD > RD^\star > 0$. Similarly, if $RD < 0$, then $RD < RD^\star < 0$.

I.5 $OR^\star = 1$ is Equivalent to $OR = 1$

It is clear from (I.5) that if $OR = 1$, then $OR^\star = 1$. Conversely, if $OR^\star = 1$, then $(1 - U)OR = 1 - U$. It follows from (I.6) that $U = 1$ if and only if $S + T = 1$, which, by assumption, is excluded. So $U \neq 1$, hence $OR = 1$.

I.6 Formula for $OR^\star/OR$ in Fig. 10.1

It follows from the identity for $OR^\star$ in (I.1) that

$$\frac{OR^\star}{OR} = \frac{[1 - T + (S + T - 1)R_1][T - (S + T - 1)R_0]}{[1 - T + (S + T - 1)R_0][T - (S + T - 1)R_1]} \times \frac{1}{OR},$$

and so $OR^\star/OR$ is a function of S, T, R_0, R_1 and OR. We have from (4.3) that

$$OR = \frac{R_1(1 - R_0)}{R_0(1 - R_1)},$$

hence

$$R_1 = \left[\left(\frac{R_0}{1 - R_0}\right)OR\right] \Big/ \left[1 + \left(\frac{R_0}{1 - R_0}\right)OR\right],$$

and so R_1 is a function of R_0 and OR. Thus, $OR^\star/OR$ is a function of S, T, R_0 and OR.

APPENDIX

Appendix for Chapter 12

J

J.1 Proof of (12.5)

Since the source population for cases and the source population for controls are the same, we have $M_1 = M_1^\dagger$ and $M_0 = M_0^\dagger$. Then

$$\widetilde{OR} = \frac{\tilde{a}_1/\tilde{b}_1}{\tilde{a}_0/\tilde{b}_0} \qquad [(12.1)]$$

$$= \frac{A_1/A_1^\dagger}{A_0/A_0^\dagger} \qquad [\text{Table 12.8}]$$

$$= \frac{[A_1/(M_1\Delta)]/[A_0/(M_0\Delta)]}{[A_1^\dagger/(M_1^\dagger\Delta)]/[A_0^\dagger/(M_0^\dagger\Delta)]}$$

$$= \frac{I_1/I_0}{I_1^\dagger/I_0^\dagger} \qquad (5.6)$$

$$= \frac{IR}{IR^\dagger}. \qquad [(5.9)]$$

APPENDIX

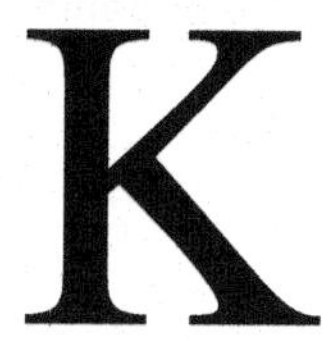

Appendix for Chapter 13

K.1 Proof of (13.1)

Tables K.1(a)–K.1(e) correspond to Tables 13.3(a)–13.3(e), respectively.

Table K.1(a) Data for closed cohort study.

Exposure	Cases	Noncases	Persons
$E=1$	R_1n_1	$(1-R_1)n_1$	n_1
$E=0$	R_0n_0	$(1-R_0)n_0$	n_0
Overall	Rn	$(1-R)n$	n

Table K.1(b) Data for cumulative case-control study embedded in closed cohort study.

Exposure	Cases	Controls
$E=1$	R_1n_1	$f(1-R_1)n_1$
$E=0$	R_0n_0	$f(1-R_0)n_0$
Overall	Rn	$f(1-R)n$

Table K.1(c) Misclassified data for cases in cumulative case-control study embedded in closed cohort study.

	$E=1$	$E=0$	Total
$E^{\star}=1$	SR_1n_1	$(1-T)R_0n_0$	$SR_1n_1+(1-T)R_0n_0$
$E^{\star}=0$	$(1-S)R_1n_1$	TR_0n_0	$(1-S)R_1n_1+TR_0n_0$
Total	R_1n_1	R_0n_0	Rn

Table K.1(d) Misclassified data for controls in cumulative case-control study embedded in closed cohort study.

	$E = 1$	$E = 0$	Total
$E^\star = 1$	$fS(1 - R_1)n_1$	$f(1 - T)(1 - R_0)n_0$	‡
$E^\star = 0$	$f(1 - S)(1 - R_1)n_1$	$fT(1 - R_0)n_0$	‡
Total	$f(1 - R_1)n_1$	$f(1 - R_0)n_0$	$f(1 - R)n$

‡ *row total.*

Table K.1(e) Misclassified data for cumulative case-control study embedded in closed cohort study.

Exposure	Cases	Controls
$E^\star = 1$	$\widetilde{a}_1^\star$	$\widetilde{b}_1^\star$
$E^\star = 0$	$\widetilde{a}_0^\star$	$\widetilde{b}_0^\star$
Overall	Rn	$f(1 - R)n$

$$\widetilde{a}_1^\star = SR_1n_1 + (1 - T)R_0n_0$$

$$\widetilde{a}_0^\star = (1 - S)R_1n_1 + TR_0n_0$$

$$\widetilde{b}_1^\star = f[S(1 - R_1)n_1 + (1 - T)(1 - R_0)n_0]$$

$$\widetilde{b}_0^\star = f[(1 - S)(1 - R_1)n_1 + T(1 - R_0)n_0]\,.$$

Setting $q = n_1/n$, we have from Table K.1(e) that

$$\widetilde{a}_1^\star = [SR_1q + (1 - T)R_0(1 - q)]n$$

$$\widetilde{a}_0^\star = [(1 - S)R_1q + TR_0(1 - q)]n$$

$$\widetilde{b}_1^\star = f[S(1 - R_1)q + (1 - T)(1 - R_0)(1 - q)]n$$

$$\widetilde{b}_0^\star = f[(1 - S)(1 - R_1)q + T(1 - R_0)(1 - q)]n\,,$$

and so

$$\begin{aligned}\widetilde{OR}^\star &= \frac{\widetilde{a}_1^\star \widetilde{b}_0^\star}{\widetilde{a}_0^\star \widetilde{b}_1^\star} \\ &= \frac{[SR_1q + (1 - T)R_0(1 - q)][(1 - S)(1 - R_1)q + T(1 - R_0)(1 - q)]}{[(1 - S)R_1q + TR_0(1 - q)][S(1 - R_1)q + (1 - T)(1 - R_0)(1 - q)]}\,. \end{aligned} \tag{K.1}$$

Computing as in the proof of (10.3) yields

$$\widetilde{OR}^{\star} = \frac{OR + U + W}{1 + (U \times OR) + W},$$

where

$$U = \frac{(1-S)(1-T)}{ST}$$

$$W = \frac{(1-S)R_1 q}{T R_0 (1-q)} + \frac{(1-T)(1-R_0)(1-q)}{S(1-R_1)q}.$$

The remainder of the proof of (13.1) is similar to the proof in Appendix I of (10.3).

K.2 $\widetilde{OR}^{\star} = 1$ is Equivalent to $OR = 1$

The proof is similar to the proof in Appendix I that $OR^{\star} = 1$ if and only if $OR = 1$.

K.3 Formula for $\widetilde{OR}^{\star}/OR$ in Fig. 13.2

By assumption, $n_1 = n_0$, hence $q = 0.5$. We have from (K.1) that

$$\frac{\widetilde{OR}^{\star}}{OR} = \frac{[SR_1 + (1-T)R_0][(1-S)(1-R_1) + T(1-R_0)]}{[(1-S)R_1 + TR_0][S(1-R_1) + (1-T)(1-R_0)]} \times \frac{1}{OR},$$

and so $\widetilde{OR}^{\star}/OR$ is a function of S, T, R_0, R_1 and OR. We have from (4.3) that

$$OR = \frac{R_1(1-R_0)}{R_0(1-R_1)},$$

hence

$$R_1 = \left[\left(\frac{R_0}{1-R_0}\right) OR\right] \Big/ \left[1 + \left(\frac{R_0}{1-R_0}\right) OR\right],$$

and so R_1 is a function of R_0 and OR. Thus, $\widetilde{OR}^{\star}/OR$ is a function of S, T, R_0 and OR.

K.4 Proof of (13.3)

Based on Tables 7.2 and 12.5, Table K.2 gives the data for stratum i of the case-cohort study. Since the subcohort is selected by simple random sampling, the sampling fraction for the subcohort in stratum i is the same (on average) as the sampling fraction f for the subcohort overall.

Table K.2 Stratum-specific data for case-cohort study.

Exposure	Cases	Subcohort
$E = 1$	$\tilde{a}_{1i} = a_{1i}$	$\tilde{b}_{1i} = f n_{1i}$
$E = 0$	$\tilde{a}_{0i} = a_{0i}$	$\tilde{b}_{0i} = f n_{0i}$

We have

$$\tilde{a}_{0(1)} = \sum_{i=0}^{k} \frac{\tilde{a}_{0i}\tilde{b}_{1i}}{\tilde{b}_{0i}} \qquad [(13.2)]$$

$$= \sum_{i=0}^{k} \frac{a_{0i} f n_{1i}}{f n_{0i}} \qquad [\text{Table K.2}]$$

$$= \sum_{i=0}^{k} \left(\frac{a_{0i}}{n_{0i}}\right) n_{1i}$$

$$= \sum_{i=0}^{k} R_{0i} n_{1i} \qquad [\text{Table 7.2}]$$

$$= a_{0(1)} \, . \qquad [(7.5)]$$

K.5 Proof of (13.4)

Based on Table 12.7, Table K.3 gives the data for stratum i of the P-B case-control study. Since the controls are selected by simple random sampling, the sampling fraction for controls in stratum i is the same (on average) as the sampling fraction f for controls overall.

Table K.3 Stratum-specific data for population-based case-control study.

Exposure	Cases	Controls
$E = 1$	$\tilde{a}_{1i} = A_{1i}$	$\tilde{b}_{1i} = f M_{1i}$
$E = 0$	$\tilde{a}_{0i} = A_{0i}$	$\tilde{b}_{0i} = f M_{0i}$

We have

$$
\begin{aligned}
\widetilde{a}_{0(1)} &= \sum_{i=0}^{k} \frac{\widetilde{a}_{0i}\widetilde{b}_{1i}}{\widetilde{b}_{0i}} && [(13.2)] \\
&= \sum_{i=0}^{k} \frac{A_{0i} f M_{1i}}{f M_{0i}} && [\text{Table K.3}] \\
&= \sum_{i=0}^{k} \left(\frac{A_{0i}}{M_{0i}}\right) M_{1i} \\
&= \sum_{i=0}^{k} I_{0i} M_{1i} \Delta && [(5.7)] \\
&= A_{0(1)}\,. && [(5.17)]
\end{aligned}
$$

K.6 $s\widetilde{OR}$ is a Weighted Average

For both case-cohort and P-B case-control studies, we have

$$
\begin{aligned}
\widetilde{a}_1 &= \sum_{i=0}^{k} \widetilde{a}_{1i} && [\text{Table 13.2}] \\
&= \sum_{i=0}^{k} \left(\frac{\widetilde{a}_{0i}\widetilde{b}_{1i}}{\widetilde{b}_{0i}}\right) \widetilde{OR}_i\,, && [(12.1)]
\end{aligned}
\qquad (K.2)
$$

hence

$$
\begin{aligned}
s\widetilde{OR} &= \frac{\widetilde{a}_1}{\widetilde{a}_{0(1)}} && [(13.5)] \\
&= \sum_{i=0}^{k} \left(\frac{\widetilde{a}_{0i}\widetilde{b}_{1i}}{\widetilde{b}_{0i}}\right) \widetilde{OR}_i \Bigg/ \sum_{i=0}^{k} \frac{\widetilde{a}_{0i}\widetilde{b}_{1i}}{\widetilde{b}_{0i}} && [(13.2), (K.2)] \\
&= \sum_{i=0}^{k} \left(\frac{\widetilde{w}_i}{\sum_{i=0}^{k} \widetilde{w}_i}\right) \widetilde{OR}_i\,. && \left[\widetilde{w}_i = \frac{\widetilde{a}_{0i}\widetilde{b}_{1i}}{\widetilde{b}_{0i}}\right]
\end{aligned}
$$

K.7 Proof of (13.6)

We have

$$
\begin{aligned}
s\widetilde{OR} &= \frac{\widetilde{a}_1}{\widetilde{a}_{0(1)}} && [(13.5)] \\
&= \frac{a_1}{a_{0(1)}} && [\text{Table 12.5, (13.3)}] \\
&= \frac{R_1 n_1}{\sum_{i=0}^{k} R_{0i} n_{1i}} && [(4.2), (7.5)] \\
&= \frac{R_1}{\sum_{i=0}^{k} \left(\frac{n_{1i}}{n_1}\right) R_{0i}} \\
&= \frac{R_1}{R_{0(1)}} && [(7.3)] \\
&= sRR. && [(7.4)]
\end{aligned}
$$

K.8 Proof of (13.7)

We have

$$
\begin{aligned}
s\widetilde{OR} &= \frac{\widetilde{a}_1}{\widetilde{a}_{0(1)}} && [(13.5)] \\
&= \frac{A_1}{A_{0(1)}} && [\text{Table 12.7, (13.4)}] \\
&= \frac{I_1 M_1 \Delta}{\sum_{i=0}^{k} I_{0i} M_{1i} \Delta} && [(5.6), (5.17)] \\
&= \frac{I_1}{\sum_{i=0}^{k} \left(\frac{M_{1i}}{M_1}\right) I_{0i}} \\
&= \frac{I_1}{I_{0(1)}} && [(5.16)] \\
&= sIR. && [(5.15)]
\end{aligned}
$$

K.9 $\widetilde{OR} \neq s\widetilde{OR}$ if and Only if $R_0 \neq R_{0(1)}$ in Case-Cohort Studies

We have

$$\begin{aligned}\frac{\tilde{a}_0\tilde{b}_1}{\tilde{b}_0} &= \frac{a_0 f n_1}{f n_0} \qquad \text{[Table 12.5]} \\ &= \left(\frac{a_0}{n_0}\right) n_1 \\ &= R_0 n_1, \qquad \text{[(4.2)]}\end{aligned} \tag{K.3}$$

and so

$$\begin{aligned}\widetilde{OR} \neq s\widetilde{OR} \quad &\text{if and only if} \quad \frac{\tilde{a}_1\tilde{b}_0}{\tilde{a}_0\tilde{b}_1} \neq \frac{\tilde{a}_1}{\tilde{a}_{0(1)}} \qquad \text{[(12.1), (13.5)]} \\ &\text{if and only if} \quad \frac{\tilde{a}_0\tilde{b}_1}{\tilde{b}_0} \neq \tilde{a}_{0(1)} \\ &\text{if and only if} \quad R_0 n_1 \neq \sum_{i=0}^{k} R_{0i} n_{1i} \qquad \text{[(7.5), (13.3), (K.3)]} \\ &\text{if and only if} \quad R_0 \neq \sum_{i=0}^{k} \left(\frac{n_{1i}}{n_1}\right) R_{0i} \\ &\text{if and only if} \quad R_0 \neq R_{0(1)}. \qquad \text{[(7.3)]}\end{aligned}$$

K.10 Proof of (13.8)

We have

$$\begin{aligned}\frac{\tilde{a}_{0i}}{\tilde{b}_{0i}} &= \frac{a_{0i}}{f n_{0i}} \qquad \text{[Table K.2]} \\ &= \frac{R_{0i}}{f}, \qquad \text{[Table 7.2]}\end{aligned}$$

and similarly

$$\frac{\tilde{a}_{0j}}{\tilde{b}_{0j}} = \frac{R_{0j}}{f},$$

hence

$$\frac{\widetilde{a}_{0i}/\widetilde{b}_{0i}}{\widetilde{a}_{0j}/\widetilde{b}_{0j}} = \frac{R_{0i}}{R_{0j}}.$$

K.11 Proof of (13.9)

We have

$$\begin{aligned}\frac{\widetilde{a}_{0i}}{\widetilde{b}_{0i}} &= \frac{A_{0i}}{f M_{0i}} && \text{[Table K.3]}\\ &= \frac{I_{0i}\Delta}{f}, && \text{[(5.10)]}\end{aligned}$$

and similarly

$$\frac{\widetilde{a}_{0j}}{\widetilde{b}_{0j}} = \frac{I_{0j}\Delta}{f},$$

hence

$$\frac{\widetilde{a}_{0i}/\widetilde{b}_{0i}}{\widetilde{a}_{0j}/\widetilde{b}_{0j}} = \frac{I_{0i}}{I_{0j}}.$$

APPENDIX

Appendix for Chapter 14

L

L.1 Proof of (14.5)

We have

$$
\begin{aligned}
\overline{y}_h &= \frac{1}{N_h}\sum_{i=1}^{N_h} y_{hi} && [(14.1)] \\
&= \frac{1}{N_h}\sum_{i=1}^{N_h}(a + bx_{hi} + c\overline{x}_h) && [(14.4)] \\
&= \frac{1}{N_h}\left(\sum_{i=1}^{N_h} a + \sum_{i=1}^{N_h} bx_{hi} + \sum_{i=1}^{N_h} c\overline{x}_h\right) \\
&= \frac{1}{N_h}\left(N_h a + b\sum_{i=1}^{N_h} x_{hi} + N_h c\overline{x}_h\right) \\
&= a + b\overline{x}_h + c\overline{x}_h && [(14.1)] \\
&= a + (b + c)\overline{x}_h\,.
\end{aligned}
$$

APPENDIX

Appendix for Chapter 15

M.1 Proof of (15.3)

Table M.1 corresponds to Table 15.1. In fact, Table M.1 is precisely Table B.1 except for the equivalent way the far-right column is expressed.

Table M.1 Population counts for screening program.

Test	$D=1$	$D=0$	Total
$D^\star=1$	SPN	$(1-T)(1-P)N$	$[1-T+(S+T-1)P)]N$
$D^\star=0$	$(1-S)PN$	$T(1-P)N$	$[T-(S+T-1)P]N$
Total	PN	$(1-P)N$	N

We have

$$\begin{aligned} PV_P &= \frac{SPN}{[1-T+(S+T-1)P)]N} \quad [(15.2), \text{Table M.1}] \\ &= \frac{SP}{1-T+(S+T-1)P} \end{aligned}$$

and

$$\begin{aligned} PV_N &= \frac{T(1-P)N}{[T-(S+T-1)P]N} \quad [(15.2), \text{Table M.1}] \\ &= \frac{T(1-P)}{T-(S+T-1)P}. \end{aligned}$$

M.2 Proof of (15.7)

Denote by S_i and T_i, respectively, the sensitivity and specificity of rater i, for $i=1,2$. Tables M.2(a)–M.2(c) correspond to Tables 15.4(a)–15.4(c), respectively. In fact, Table M.2(c) is precisely Table 15.3.

Table M.2(a) Rater assignments of subjects in DPCP for reliability study.

Rater 1	Rater 2		Total
	$D^\star=1$	$D^\star=0$	
$D^\star=1$	S_1S_2Pn	$S_1(1-S_2)Pn$	S_1Pn
$D^\star=0$	$(1-S_1)S_2Pn$	$(1-S_1)(1-S_2)Pn$	$(1-S_1)Pn$
Total	S_2Pn	$(1-S_2)Pn$	Pn

Table M.2(b) Rater assignments of subjects not in DPCP for reliability study.

Rater 1	Rater 2		Total
	$D^\star=1$	$D^\star=0$	
$D^\star=1$	$(1-T_1)(1-T_2)(1-P)n$	$(1-T_1)T_2(1-P)n$	$(1-T_1)(1-P)n$
$D^\star=0$	$T_1(1-T_2)(1-P)n$	$T_1T_2(1-P)n$	$T_1(1-P)n$
Total	$(1-T_2)(1-P)n$	$T_2(1-P)n$	$(1-P)n$

Table M.2(c) Rater assignments for reliability study.

Rater 1	Rater 2		Total
	$D^\star=1$	$D^\star=0$	
$D^\star=1$	a_{11}	a_{10}	$a_{11}+a_{10}$
$D^\star=0$	a_{01}	a_{00}	$a_{01}+a_{00}$
Total	$a_{11}+a_{01}$	$a_{10}+a_{00}$	n

$$a_{11}=[S_1S_2P+(1-T_1)(1-T_2)(1-P)]n$$

$$a_{01}=[(1-S_1)S_2P+T_1(1-T_2)(1-P)]n$$

$$a_{10}=[S_1(1-S_2)P+(1-T_1)T_2(1-P)]n$$

$$a_{00}=[(1-S_1)(1-S_2)P+T_1T_2(1-P)]n$$

$$a_{11}+a_{01}=[1-T_2+(S_2+T_2-1)P]n$$

$$a_{10}+a_{00}=[T_2-(S_2+T_2-1)P]n$$

$$a_{11}+a_{10}=[1-T_1+(S_1+T_1-1)P]n$$

$$a_{01}+a_{00}=[T_1-(S_1+T_1-1)P]n\,.$$

Denote by $P_i^\star$ the (misclassified) point prevalence based on rater i, for $i = 1, 2$. Corresponding to (3.5) is

$$P_i^\star = 1 - T_i + (S_i + T_i - 1)P\,,$$

from which it follows that

$$\begin{aligned} a_{11} + a_{01} &= P_2^\star n \\ a_{10} + a_{00} &= (1 - P_2^\star)n \\ a_{11} + a_{10} &= P_1^\star n \\ a_{01} + a_{00} &= (1 - P_1^\star)n\,. \end{aligned}$$

We have from (15.4) and (15.5) that

$$\begin{aligned} p_{\text{agree}} &= \frac{a_{11} + a_{00}}{n} \\ &= [S_1 S_2 + (1 - S_1)(1 - S_2)]P \\ &\quad + [T_1 T_2 + (1 - T_1)(1 - T_2)](1 - P) \end{aligned} \tag{M.1}$$

and

$$\begin{aligned} p_{\text{chance}} &= \frac{(a_{11} + a_{01})(a_{11} + a_{10}) + (a_{10} + a_{00})(a_{01} + a_{00})}{n^2} \\ &= [1 - T_1 + (S_1 + T_1 - 1)P][1 - T_2 + (S_2 + T_2 - 1)P] \\ &\quad + [T_1 - (S_1 + T_1 - 1)P][T_2 - (S_2 + T_2 - 1)P]\,. \end{aligned} \tag{M.2}$$

Lengthy but straightforward computations using (M.1) and (M.2) yield

$$p_{\text{agree}} - p_{\text{chance}} = 2P(1 - P)(S_1 + T_1 - 1)(S_2 + T_2 - 1) \tag{M.3}$$

and

$$\begin{aligned} 1 - p_{\text{chance}} &= [1 - T_1 + (S_1 + T_1 - 1)P][T_2 - (S_2 + T_2 - 1)P] \\ &\quad + [T_1 - (S_1 + T_1 - 1)P][1 - T_2 + (S_2 + T_2 - 1)P]\,. \end{aligned} \tag{M.4}$$

Setting $S_1 = S_2 = S$ and $T_1 = T_2 = T$ in (M.3) and (M.4) and substituting into (15.6) gives (15.7).

M.3 Raters are Perfectly Valid or Perfectly Invalid

If the rater-screening test combinations are perfectly valid, that is, if $S_1 = S_2 = 1$ and $T_1 = T_2 = 1$, then (M.1) gives $p_{\text{agree}} = 1$. On the other hand, if the rater-screening test combinations are perfectly invalid, that is, if $S_1 = S_2 = 0$ and $T_1 = T_2 = 0$, then again (M.1) gives $p_{\text{agree}} = 1$.

Index

S

T

U

V

W

Printed in the United States
by Baker & Taylor Publisher Services